The British Museum (NATURAL HISTORY)

Text by PETER WHITEHEAD
Photographs by COLIN KEATES

Philip Wilson

*Published in association with
the British Museum (Natural History)*

© Text and illustrations British Museum (Natural History) 1981

First published 1981 by
Summerfield Press Ltd and Philip Wilson Publishers Ltd
Russell Chambers, Covent Garden, London WC2E 8AA

Text: Peter Whitehead
Photographs: Colin Keates
Design: Paul Sharp
Edited by Philip Wilson Publishers Ltd

Produced by Scala Istituto Fotografico Editoriale, Firenze
Printed in Italy

ISBN 0-85667-108-8 Pbk
ISBN 0-85667-109-6 Hdbk

Front cover: Blue and yellow macaw *(Ara ararauna)*
of South America, from a watercolour
in the Reeves Collection.

Contents

Foreword

This book highlights some of the treasures in the collection of the Natural History Museum, recounts its history, and describes the scope of the scientific work going on behind the scenes. The account is inevitably brief and selective, but I hope that readers will be fascinated and delighted by this reflection of a unique national museum.

The British Museum (Natural History) has been open to the public in South Kensington for one hundred years, and each year the public sees the results of one aspect of its function—the exhibition and educational activities. In our public galleries we aim to provide a stimulating environment where visitors can see and learn something of the diversity and interrelationships of the world's natural objects, and of the scope and achievements of modern natural history in its broadest sense. We were delighted to be chosen National Heritage 'Museum of the Year' for 1980. This award, sponsored by *The Illustrated London News* in conjunction with National Heritage, was given in recognition of exhibition and educational activities over the period 1977–9.

By far the major part of the Museum's activities, however, relates to its scientific function as one of the world's leading taxonomic institutes. Taxonomy is concerned with the identification and classification of animals, plants, fossils, and minerals, of which the Museum has a vast collection of reference specimens. This priceless national collection forms an indispensable basis for our scientific work in support of agricultural, medical, geological, and other environmental sciences. Knowledge of the natural world has never been so important as it is today, and research associated with the Museum's collections plays an important role in extending this knowledge, as this book illustrates.

The text was written by Dr Peter Whitehead, a zoologist on the Museum's scientific staff, and the photographs were taken by Colin Keates of our Photographic Unit. Dr. Gordon Corbet, Head of the Museum's Central Services Department has been involved with the book from its inception and has kindly co-ordinated its preparation.

R. H. Hedley
DIRECTOR
June 1980

4

Introduction

On 18 April, the Easter Monday of 1881, the first visitors filed into the newly built natural history museum in South Kensington, a part of London that was already an important site for exhibitions in the arts and sciences. The new museum was given the title British Museum (Natural History), although it is generally called the Natural History Museum; to research workers it is just 'the BM'. With the sapling plane trees grown to majestic height and the imposing façade cleaned of a century's grime, it has become one of the most popular of all London's museums. Up to three million visitors pass through its doors annually, while in the field of research its reputation is immense. Its founders obviously hoped for such things and their successors strived with great energy, but above all this growth stems from the fact that nowadays natural history is seen to have a great deal to do with everyday life.

In its time the Museum has been host to some five generations of children and indeed the most frequent remarks when the Museum is mentioned are 'Oh yes, I remember going there as a child' or 'I've been meaning to take the children again'. Of course, the Museum is not just a children's museum, but there is no doubt of its popularity with them and no doubt either that its image is stamped early on their general conceptions about natural history. It joins other images, of zoos, botanical gardens, television programmes, books, and real encounters with the natural world, to shape attitudes towards some of the most pressing problems of our times—the conservation, use, and management of our natural resources. In this broad context, and long before the precise names of animals, plants, and minerals are learned and their life processes grasped, the Museum makes its almost unseen mark on our social response to nature.

In 1881 the Victorians who conceived the Museum were no less aware of their responsibility, but a century has achieved new insights and provoked new goals. Nevertheless, the basic functions of the Museum are the same: not just to look after and enhance the national collections, but to educate and to carry forward research in the broad field of natural history. Each stimulates the other, on the one hand helping to prevent research from hiding in some 'ivory tower', and on the other keeping the public who visit aware of new developments and ideas.

The public galleries hardly need description—or they will have failed in their intention. However, the research collections, and more importantly the way that they are used, are not at all obvious to the visitor and people are sometimes surprised that anything more than administration goes on behind all those locked doors. An earnest young man appears at one door, a learned sage disappears through another, but exactly what they do is mysterious. Yet the Museum as a research as well as an educational institution is supported by public funds and the scope and relevance of its research activities should be more generally known and, we hope, approved.

In its role as a research institute the Museum is essentially a vast storehouse of some 50 million specimens of animals, plants, fossils, and minerals. Its research on these is basically the discipline known as taxonomy, or the attempt to identify, name, and classify the natural world, which is the first step towards understanding its processes and being able to use it. A cow is plainly a cow and a giraffe hardly to be confused with anything else, but with 4000 species of aphid and 14 000 species of ants, how do we recognize the harmful, the harmless, and the useful? If one is harmful in Europe, can we suppose that its very close relative in Africa is also harmful? All biological work needs this basis of classification and the Museum serves as a centre for such studies, with its vast collections almost unrivalled and certainly the finest of their kind in very many groups.

This book presents a selection of the many and varied activities that go on everyday behind the locked doors (but doors which are gladly opened to thousands of scientific or other visitors seeking further information or wishing to study the specimens or literature). Rare, unusual, or striking specimens are shown alongside apparently mundane objects, the latter often with far more scientific or social consequences than their more colourful neighbours. The story of the collections and of their size and scope, as well as the varied and often surprising uses to which they are put, is an important chapter in the general development of the natural sciences. It gives to the Museum a roundness of purpose in its attempts to serve society in the field of natural history.

Sir Hans Sloane bequeathed his enormous museum and library to the nation in 1753 to become the nucleus of the British Museum. Born at Killyleagh, County Down, Sloane came to London to study medicine, continuing his studies in France and then entering the practice of the famous London physician Thomas Sydenham. Two years later Sloane made his celebrated voyage to Jamaica, which resulted in his catalogue of Jamaican plants (1696) and also his natural history of Jamaica (1707–25). An eminent physician and keen naturalist, he rose to be President of both the Royal Society and Royal College of Physicians. He was an avid collector and some 80 000 of his specimens were incorporated into the British Museum; some of these are still preserved at South Kensington, in particular his large herbarium. Sloane Square and Hans Crescent were named in his honour.

Origins of the Museum

Museums contain the working material of those engaged in classifying objects, for this is always the first step in any investigation, the assembling of the data; objects only become significant when placed, either physically or theoretically, against other objects and the differences noted. Thus a museum is a 'classifying house'. The classification underlying a natural history museum is more subtle than most, for it depends, ultimately, on the interrelationships between individuals, populations, species, and their environments—on the whole spectrum of interrelations that have woven the pattern of evolution and produced the multitude of plant and animal forms.

The word 'museum' originally stemmed from the Greek word describing a place dedicated to the Muses and therefore suitable for learned discussion and study (we still 'muse' on a topic). In antiquity, such places may well have kept objects for study, and certainly Aristotle and Pliny must have had private collections, but the modern museum effectively dates from the Renaissance when men of wealth kept cabinets of 'curiosities' to show to their friends. Even at that time, however, one can already see the dual role of museums: to exhibit objects and to provide a working collection for scholars. The first natural history museum was perhaps that of Conrad Gessner (1516–65) of Zurich, one of the great encyclopedic writers of the period. Essentially, Gessner was a classifier and his museum and library were his working tools. Nowadays the word *taxonomy* is used for the classifying of animals and plants (from the Greek *taxis* or arrangement, hence also taxidermy or the arrangement of skins). Taxonomy requires institutes where thousands or even millions of specimens are stored for comparison and study.

The early museums were privately owned or were the museums of learned societies and they were intended for friends or for fellow scholars. The first public museum in England was that of John Tradescant (1587–1638) of South Lambeth, early in the seventeenth century. It was extremely popular and was dubbed 'Tradescant's Ark', but it was not until the end of the eighteenth century that the really large public museums arose. The Leverian Museum and Bullock's Museum at the Egyptian Hall, both in London, were pioneers in the use of museums to educate people, but they were expensive to run and both Lever and Bullock were forced to sell up. They had shown that museums were popular, but they had also shown that large museums must be the responsibility of a tax-paying public.

It is significant that the present Museum in South Kensington owes its origin to a private collection, that of Sir Hans Sloane (1660–1753). Born at Killyleagh, County Down, in the year of the foundation of the Royal Society (of which he would later be President), Sloane took up medicine, later travelling to Jamaica as physician to the Duke of Albemarle before settling down to a highly successful medical career in London. An avid, even eccentric collector, he built up an enormous museum and library at his home in Chelsea, the manor house built by Henry VIII. 'I am but just come from Sir Hans Sloane's,' the Duchess of Portland once remarked, 'where I have beheld many odder things than himself, though none so inconsistent.' When Sloane died, at the age of ninety-two, he bequeathed this vast and 'odd' collection to the nation on condition that his two daughters together received £20 000 (perhaps a quarter of its real value) and he asked that powers be granted 'for continuing and preserving my said collection, in such manner as . . . most likely to answer the public benefit'. By an Act of Parliament, passed three months after Sloane's death, the purchase was agreed together with that of the Harleian Collection of manuscripts, the two to be combined with the Cottonian Library in one general repository. Money was raised by lottery and the collections were installed in Montagu House in Bloomsbury. On 15 January 1759 the public first entered the new British Museum.

Sloane's 'curiosities', which comprised something like eighty thousand items, became the Department of Artificial and Natural Productions (the other two Departments being for books and for manuscripts). Virtually all objects were on display, but visitors were whisked through the rooms of Montagu House in parties 'with leisure just to cast one poor longing look of astonishment on all these stupendous treasures of natural curiosities, antiquities and literature' as one disgruntled foreigner complained. James Empson, Sloane's former curator, looked after this conglomeration of objects. By 1837, natural history had three Branches (Botany, Zoology, and Mineralogy with Geology) and by 1857 these had become four Departments under Richard Owen (1804–92), who was later to be the director (in all but name) of the present museum in South Kensington. For some seventy years the collections were crammed into Montagu House until the erection of the neo-Greek British Museum building just behind it. Poor condemned Montagu House, as Charles Lamb called it, was finally pulled down in 1849 after the gradual transfer of the natural history material to the new British Museum.

For a time there was more space for exhibiting and working on the natural history and other collections, but it is worth remembering the conditions under which taxonomy was carried out in those early days. In 1877, Philip Sclater, the Secretary to the Zoological Society, gave a visitor's impression of the Zoology Department:

... descending (with care) a flight of darkened steps, he will find himself in the cellar ... Two small studies partitioned off to the left are assigned to the keeper of the department and his assistant. The remaining naturalists are herded together in one appartment commonly called the 'Insect-room' ... No lights are allowed, and when the fogs of winter set in, the obscurity is such that it is difficult to see any object requiring minute examination.

Nearly twenty years earlier it had been decided that 'all the Natural History Collections be speedily and simultaneously removed', but there were to be endless delays. Eventually, the present site in South Kensington was bought and, as a result of a competition, the design of the new museum was entrusted to Francis Fowke (1823–65), architect of the Royal Scottish Museum in Edinburgh. On Fowke's early death, the task was given to Alfred Waterhouse (1830–1905), already well known as architect of the Manchester Assize Courts. Although influenced by Ruskin and the revival of Gothic, Waterhouse chose as his theme the Romanesque style of the eleventh and twelfth centuries and decided on a complete façade of terracotta in beige and grey–blue. Side-wings, postponed for lack of funds, were never built, but with its cathedral-like appearance, it still made a most impressive storehouse for the 'Wonders of Creation'.

The building was completed in 1880 and on Easter Monday the following year it was opened to the public. It had taken seven years to build, had cost £412 000, and was the most impressive of all Waterhouse's buildings. Mineralogy was moved in first, followed by Botany and Geology, and finally by Zoology in 1882–3. Richard Owen, Superintendent of the four Departments, was now in his eightieth year and with some satisfaction retired in December 1883. His sketches, drawn as early as 1859 and 1862, had served as the basis for the final design (although greatly modified) and it had been his energy and drive that had largely brought the new Museum into being. He was succeeded by William Flower (1831–99), a most able museum man, and the Museum embarked on a new phase of its development.

In just over a century before the move to South Kensington, the Natural History Departments of the British Museum had built up an enviable reputation. As the collections grew in the nineteenth century there came the great phase of cataloguing, initially to assist the visitor, but increasingly to make the non-exhibited specimens known to outside workers. Inevitably the catalogues became scientific works, revisions of the classification of groups of animals, plants, and minerals, and in this way the staff evolved from mere custodians of the material to active scientists. As a relict of former times, however, the heads of the scientific departments are still called Keepers.

Once free from the crippling restrictions of space, the new Museum began to expand and diversify. In 1913 a separate Department of Entomology was split off from Zoology, making five research Departments in all (Geology was designated Palaeontology in 1956). In recent years four more Departments have been created: Library Services, Central (technical) Services, Public Services, and Administrative Services. Lord Rothschild's private natural history museum at Tring in Hertfordshire was bequeathed to the Museum in 1937 and in 1971 became the home of the Museum's Sub-Department of Ornithology. Between 1920 and 1930 a New Spirit Building was constructed for storage of the specimens in alcohol and to provide laboratory space. In 1936 the Department of Entomology occupied the first half of its present block, which was to be completed in 1952. New wings accommodate the mammal collections, the General Library and Lecture Theatre (omitted from Waterhouse's plan), the exhibition staff, and the administration, while to the visitor the most obvious addition is the modern eastern extension of the main façade which houses the Department of Palaeontology. In 1963 the Museum became completely independent of the British Museum by being granted its own governing body of twelve Trustees. Two years later, responsibility for its finances was transferred from the Treasury to the Department of Education and Science.

Sir Hans Sloane might well feel bewildered by the complexity of the present Museum, but he would probably acknowledge it as a reflection of the increasing complexity of science and society. Although his pious hope, that his collections should serve to confute 'atheism and its consequences', may wilt before Thomas Huxley's statue in the North Hall, the Museum together with its parent body in Bloomsbury are still firmly dedicated to the 'use and improvement of the arts and sciences, and benefit of mankind'.

Sir Richard Owen was the controversial Superintendent of the Natural History Departments from 1856 until his retirement at the end of 1883. His superb intellect and quite remarkable work as a comparative anatomist and palaeontologist were, according to his contemporaries, marred by jealous and devious ways. Nevertheless, it was largely through his energy and drive that the present Museum in South Kensington was built. Trained in the medical profession, he had been Conservator at the Hunterian Museum of the Royal College of Surgeons in London until his transfer to the British Museum. In 1839 he was shown a small piece of bone found in New Zealand and from this apparently slender evidence he deduced that it was part of the femur of a large, ostrich-like bird (it was in fact from an extinct moa). Four years later he received many more bones, which proved his deduction correct. In this statue he holds a bone, probably not the fragment of moa femur, but certainly a symbol of his great skill as an anatomist.

9

This water-colour, by an unknown artist, shows Montagu House in Bloomsbury, the first home of the British Museum. The original Montagu House had burned down in 1685, but it was rebuilt in about 1700 by Ralph, first Duke of Montagu to the design of Peter Puget from Marseilles. In his will, Sir Hans Sloane had wanted his collections to stay in his home, the Manor House in Chelsea, but the newly appointed Trustees of the Museum decided otherwise. On 15 January 1759 the public first entered the national museum and were conducted round the exhibits. A magnificent staircase, with a stuffed rhinoceros and three giraffes at the top, led to the natural history rooms on the first floor. Montagu House was used for some seventy years, until completion of Sir Robert Smirke's new building just behind it, the classical façade of which is still a most impressive sight.

Owing to the early death of Francis Fowke, who had been commissioned to design the new building for the Natural History Departments of the British Museum, his plans were handed over in 1868 to Alfred Waterhouse. Waterhouse, like many architects of the day, was much influenced by Ruskin, who had held that 'whatever you really and seriously want, Gothic will do it for you'. Waterhouse, however, decided that the earlier round-arched Romanesque style could do it better in view of the numerous terracotta decorations requested by

Owen. His original design had wings at the two ends, but for reasons of economy these were omitted (and were never built). The huge façade, shown here in an original drawing by Waterhouse, is 205 metres long and the two central towers are 52 metres high. From the air one can see that the towers of the Museum are aligned with the Albert Memorial and Royal Albert Hall (with the tower of Imperial College and those of the Royal College of Music on this same line).

Exactly a hundred years ago the Keeper and staff of the Geology Department posed for this photograph at the end of the colonnade at the back of the newly built Museum. Standing there today, they would see Keepers of Departments in sports jackets, research workers with open collars, and assistants wearing jeans, but they would surely be pleased to know that their work has been continued with just as much seriousness as in their day. They would also be considerably surprised at the enormous growth of the collections and of the staff, as well as the extraordinary sophistication of the equipment now used. From left to right (standing) Arthur Smith Woodward, William Davies and R. Bullen Newton, Assistants; (seated) Robert Etheridge, Assistant Keeper, and Henry Woodward, Keeper.

The new Palaeontology wing, extending to the east of the main Museum frontage, was formally opened on 24 May 1977 by Mrs Shirley Williams, Secretary of State for Education and Science. Fully air-conditioned, it was designed by architects of the Department of the Environment in conjunction with consultant architect John Pinckheard, with instructions to meet the very special needs of the Department of Palaeontology. It has an 'open plan' design and provides 10 000 square metres of floor area on seven floors for the study and storage of fossil specimens. The architectural press signalled its approval for the way that the building retains its own integrity yet harmonizes with the Victorian Romanesque style of the main building, as well as with the neo-Palladian façade of the Geological Museum which it abuts.

11

John Edward Gray (1800–75) was one of the most notable figures in the early history of zoology at the British Museum. Beginning with a temporary appointment at 15 shillings a day in 1824, he rose to be Keeper of Zoology in 1840 and on his retirement only two months before his death he had devoted fifty years to the Museum. Never a specialist, he published a thousand papers on a variety of animals and wrote half of the two hundred catalogues that were published at the Museum during his time. His interests were exceptionally wide and outside of scientific circles he was best known for his *Hand catalogue of postage stamps*; he claimed that it was he, together with Rowland Hill, who laid the foundations of penny postage.

This early water-colour captures something of the elegance with which the finely made display cases set off the interior of Waterhouse's superb design. Although the displays were often crowded with specimens, the high ceilings gave a sense of space to the galleries while the architectural details, and especially the terracotta motifs, encouraged the eye to wander from time to time so that the visitor could reflect on the exhibits. The polished wooden cases blended well with the beige terracotta, thus allowing the objects themselves to provide the visual excitement. By modern standards, the information given on the labels was rather sparse; the name was supplied and perhaps a note on its occurrence, habits, or uses, but thereafter the visitor was expected to relate it to something he had read or to be sufficiently interested that he would go home and try to find out. Not everyone would have responded in this way, but few can have emerged from this cathedral-like building without a sense of awe at the complexity of the world around them, largely as a result of the setting that Waterhouse provided.

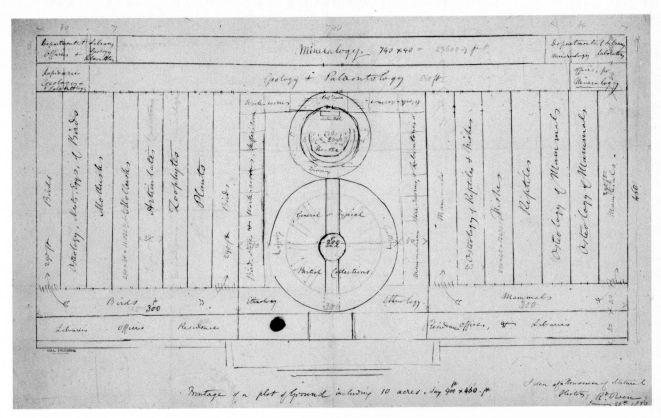

Richard Owen became Superintendent of the Natural History Department in 1856 at a time when space for the collections and exhibitions at the British Museum in Bloomsbury was crippling. In 1858, as President of the British Association for the Advancement of Science, he aired his views and in January the following year he sketched out the plan for a new natural history museum shown here. Central to his concept and to the building was to be a circular area for a lecture theatre and for his cherished idea of an 'index museum' epitomizing the main types of animals, plants, and minerals. In a second plan, drawn up in 1862, he halved the size of the plot required from 10 acres to 5 (about 2 hectares). Owen's plans served as guides for the first design by Francis Fowke and for the final design by Alfred Waterhouse, but in some mysterious way the lecture theatre disappeared.

Sir Joseph Banks (1743–1820), who as a young man had served as naturalist on the first of Captain Cook's voyages around the world, later became a Trustee of the British Museum and was one of the important donors of natural history specimens. Through Banks the Museum acquired a very large collection of natural history drawings as well as plants and some animals from all three of the Cook voyages, and from other expeditions also. His interests were wide and although he made no direct contribution to natural history he acted as a patron and catalyst, bringing people together, helping to make their results known, and always keeping a keen eye on the practical application of new discoveries in botany and zoology. His home in Soho Square served as a virtual museum and meeting place for naturalists from all over the world. Like Sloane, he rose to be President of the Royal Society, a post he held for no less than forty years. This statue, by Sir Francis Chantry, used to stand in the entrance of Montagu House, the first home of the British Museum; it is now outside the Botany Gallery.

The Main Hall of the Museum was originally intended to contain the 'index museum', an idea cherished by Richard Owen. In six bays on one side would be minerals, plants, and invertebrate animals, and on the other side would be the vertebrates, the whole forming a simple guide to the principal 'types' of the animal, plant, and mineral kingdoms. In the centre Owen wanted examples of the largest animals, including the blue whale, elephants, and giraffes. Owen's successor, William Flower, introduced the theme of evolution (to which Owen had been strongly opposed). In 1958, to coincide with the centenary of Darwin's first paper on the origin of species by natural selection, the bays were given over entirely to evolutionary themes. As a last hint of Owen's ideas, however, the central area displayed representatives of zoology, entomology, botany, palaeontology, and minerals, together with elephants.

For the decoration of the Museum, Waterhouse chose beige and grey-blue terracotta and one of the most charming aspects of his design is the terracotta menagerie, both outside and inside. For the west wing he chose living animals and for the east wing extinct animals. Although he showed early talent as an artist, his Quaker family apparently thought that architecture was more respectable. His skill as a draughtsman and indeed his interest in drawing is shown in the sketches that he made for the terracotta animals. The subjects were mainly chosen by Richard Owen but Waterhouse personally designed all of them. Dujardin, a French modeller, produced casts in plaster, from which the terracotta ones were made by Messrs. Gibbs & Canning of Tamworth in Staffordshire. Among such familiar subjects as the red squirrel shown here is the passenger pigeon, which in 1881 was still quite common in North America, but became extinct in 1914.

COFFEA·ARABICA   NICOTIANA·TABACVM   GOSSYPIVM·BARBADENSE

From the stairs at the end of the Main Hall the statue of Richard Owen now looks down, appropriately enough, on a display of dinosaurs, a name that he himself had proposed in 1841 for these extinct and often enormous reptiles. The Main Hall is 52 metres long and almost half as high. Dominating this great space is the longest of all dinosaurs, *Diplodocus carnegiei*. The 26-metre skeleton, cast from the original specimen at the Carnegie Museum in Pittsburgh, was given to the Museum in 1905 by the Scottish–American industrialist Andrew Carnegie.

The dominant theme for the terracotta decorations is zoological, but the ceiling of the Main Hall is given over entirely to plants. There are nine bays, each with 18 panels, or a total of 162 panels. The designs are large and bold, for more detailed work would have been lost in a hall that is 52 metres long and some 22 metres high. The panels on either side of the ridge are unlabelled and highly stylized, most being difficult to identify. The two rows below them are more accurate and the name of the species is written on the beam below. In the main part of the hall there are twelve of these species, each occupying a block of six panels, while over the stairs at the southern end there are 54, each species on a separate panel. The plants are of economic or medicinal importance, such as apple, lemon, maize, aloe, tea, coffee, castor-oil, tobacco, and so on. The panels were painted by Messrs Best & Lea of John Dalton Street, Manchester.

Lionel Walter Rothschild (1868–1937), second Baron Rothschild, was one of the great patrons of natural history. Born into a great banking family, he had a life-long interest in collecting and studying animals and he used his position and considerable wealth to promote not only his own work but that of others. For his museum at Tring in Hertfordshire he employed two very able and scientifically distinguished curators and the three of them published more than 1700 scientific books and papers and described more than 5000 new species of animals. Rothschild's main interest was in insects and birds and one of his best-known works was *Extinct birds*, published in 1907. Another of his interests was in the giant tortoises of Aldabra and the Galapagos Islands and at one time he even bought Aldabra Island in the Indian Ocean to save its tortoises from extinction.

Lord Rothschild's interest in natural history began early. At sixteen he met and began corresponding with Albert Günther, Keeper of Zoology at the Museum and at twenty-one his father gave him land on the outskirts of Tring Park, in Hertfordshire, where he built two small cottages, one for his books and insects, the other for a caretaker; behind this he built a larger house for his growing collections of mounted specimens. He opened his museum to the public in 1892. Outbuildings were erected as the collection grew, to be replaced by new wings to the main building in 1908–12. It was through Günther that Ernst Hartert was employed by Lord Rothschild as a curator, to be joined by Karl Jordan shortly after, and it was Günther again who suggested that the Tring Museum issue its own scientific journal, the *Novitates zoologicae*. In his will, Lord Rothschild bequeathed the entire museum to the Trustees of the British Museum, provided that it became an annexe of the Museum at South Kensington and continued as a centre for zoological research. In 1971, the Sub-Department of Ornithology and the national collection of birds were moved to Tring, where they are housed in a modern extension.

More than a hundred thousand people visit the Zoological Museum at Tring each year. The buildings and the setting have great charm and the interior retains something of the flavour of the great private museums that flourished in the last century. By the 1970s, parts of it were excessively overcrowded with specimens and a modernization programme was begun. The show cases were given better lighting and in some instances reorganized, the older specimens were restored, others were removed, and more informative labels were added. Wherever possible, however, the character and general arrangement were retained. When Lord Rothschild bequeathed it to the nation in 1937, it was the largest collection of natural history specimens ever assembled by one man; of butterflies and moths there were some two and a half million, of bird skins three hundred thousand, and the collections of other animals were equally vast.

Conrad Gessner (1516–65) of Zurich was one of five outstanding scientific men who revolutionized the study of natural history, and in particular zoology, during the Renaissance, the others being Rondelet, Salviani, Belon, and Aldrovandi (all born within 15 years of each other). At the same time, botany also had its 'fathers'—Otto Brunfels, Hieronymus Bock (Tragus), Leonhart Fuchs, and Valerius Cordus, all of whom were German. By the end of the sixteenth century these men had lifted natural history out of its mediaeval trappings and slavish reliance on ancient Greek writings and had provided a basis for the modern exploration of the natural world. Four volumes of Gessner's encyclopaedic *Historia animalium* were published in his lifetime, but he is also remembered for having probably the first museum largely devoted to natural history; four centuries later, museums play an essential role in the study of plants and animals, but the days of the privately owned museum are virtually gone.

The four-year voyage of HMS *Challenger* (1872–6) was in
many ways the birth of modern oceanography. Among the
very many results, published afterwards in fifty volumes, was
the proof that the primitive 'life substance' on the ocean
floor, which Thomas Huxley had rashly named *Bathybius*,
was nothing more than calcium sulphate precipitated by the
alcohol in which the sample was kept. The origin of life was
not in the depths. The *Challenger* collection of oceanic
sediments, manganese nodules, and rocks was donated to the
Museum in 1922 and forms the nucleus of the Ocean Bottom
Deposits Collection. More than a century later, requests are
often received to re-examine this historic material.

Growth of the collections

The founding collections of the British Museum were those of Sir Hans Sloane, together with the large Cottonian Library and the Harleian Collection of manuscripts. Sloane's museum, on which he is said to have spent between £50 000 and £100 000 (about 33 times as much in modern purchasing power), was a vast assemblage of plant, animal, mineral, and fossil specimens, antiquities, artefacts, books, and manuscripts, not all of which had much relevance even to the science of his day. In 1691 the diarist John Evelyn saw the collection at Sloane's Manor House in Chelsea and found it 'very copious and extraordinary'. The composer Handel came to tea, but carelessly laid his buttered muffin on a valuable book, 'which put the old bookworm terribly out of sorts' as the story goes. More important was the visit of the Prince of Wales in 1748, who seems to have hinted that the collection should one day be the property of the nation. Possibly Sloane had already decided this, for his two daughters could hardly have had much use for it.

The Sloane collection comprised about 50 000 books and 79 575 'objects', over half of which were in some way connected with natural history. There were 1886 mammals, 1172 birds (or their eggs or nests), 1555 fishes, 5439 insects and, of great importance, his 334 herbarium volumes containing thousands of sheets of pressed plants; there was also a large number of minerals, rocks, and fossils. It is sad to relate that all too little of this has survived. Some of the lost specimens would be of general historical value and interest, while others could perhaps throw light on early domestic animal breeds, on dates of introductions of animal species, on the variation of certain species after nearly 300 years, or on the identity of species later named on the basis of Sloane's material or descriptions. Fortunately, Sloane's herbarium has survived and is a fund of botanical information.

The loss of so many of Sloane's original specimens reflects partly a diminishing interest in the type of material resulting from his somewhat indiscriminate taste, and partly the difficulties of preserving specimens in those days. The skins of mammals and birds were prone to damage from insects (which could also destroy insect collections), while preservation in alcohol (in use since the 1660s) required very carefully sealed jars if the liquid was not to evaporate in a few years. Sloane's curator, James Empson, did his best, but George Shaw (1757–1813), the Keeper of the Zoology Department from 1806, combined neglect with the destruction of its inevitable results, large amounts of Sloane material being consigned to what he jocularly called his annual 'cremations'.

Meanwhile, the collections at the British Museum were beginning to grow as a result of British interests overseas. The first trickle from the voyages of Captain Cook (1768–80) and from voyages that succeeded them was later augmented by the considerable amount of Cook and other material donated by Sir Joseph Banks, although as a Trustee of the Museum he advised against the wholesale purchase of some of the larger private museums. By the mid-nineteenth century the influx of material was enormous. The early registers of specimens, begun in the present form in 1837, show collections of thousands of specimens which the badly under-staffed Museum had to identify, label, and incorporate. In Zoology alone, some 200 000 specimens were added between 1856 and 1861, and in the next eight years another 300 000 were acquired, mostly by donation.

It was Banks who had pioneered the custom of including a naturalist on voyages of discovery, but a century later came voyages such as that of HMS *Challenger* (1872–6) which were made for purely scientific reasons, and the quantity of specimens collected was thereby increased. The growth of the Empire, the popularity of big-game hunting, the increasing numbers who turned to natural history as a leisure activity (mostly based on collecting), the interest in horticulture—all these and many other factors, including the growing prestige of the Museum itself, ensured that the staff were kept busy. Among the early acquisitions were specimens from Darwin's voyage on the *Beagle*, Alfred Wallace's insects and birds, virtually the entire museums of the Zoological Society and of the East India Company, Brian Hodgson's birds from India and Nepal, the 80 000 insects of J. C. Bowring's collection and a similar number of shells from Hugh Cuming's, the superb collection of the Hon. Charles Greville's minerals, and the highly important herbarium that had belonged to Banks. It is no wonder that by 1860 the Museum was bursting at its seams. All this took place at a time when it was believed that a species could be diagnosed on the basis of just a few specimens. Nowadays, when the true nature of variation is investigated through large series of specimens, the growth of the collections is simply enormous, amounting to as much as half a million additions each year. This is in spite of the fact that the days of indiscriminate collecting during the exploration of new territory are long past. Collecting is now increasingly done by the Museum's own research workers, who can be highly selective in what they preserve, concentrating for example on groups such as aphids, soil mites, diatoms, or microscopic fossils that received scant attention in earlier times, and devoting as much care to the documentation of the environmental conditions in *which the animals or plants were living as to the collection of the specimens themselves.*

One might ask why such enormous collections are needed. After all, once a species has been recognized and described, why keep more than a few samples of it? Theoretically, this is what the Museum has done, for with about 2 million species of animals, plants, and minerals known, and approaching 50 million specimens in the collections, this makes only about 25 of each. In fact, the real situation is quite different. Very many species of plants and animals (at least a third, probably more) are either not represented at all or perhaps by only one or two specimens, which may anyway be in poor condition. Some of these are indeed rare species, restricted to some small locality, but more often they are common enough where they live but few people have collected from the area; a botanist on a two-week collecting visit to Guatemala may bring back an interesting frog for his colleague, but he is quite busy enough with his own speciality. As a result, the Museum's collections are frequently patchy and very often represent the interests of successive generations of specialists.

The need for more than just a few specimens from each species stems from the nature of the work. Thus, a great many more names have been given to species than is actually justified, either because the describer did not fully study the literature to see if his species was truly a new one, or because the early descriptions were not adequate. Again, the range of variation between individuals (or between the sexes or the growth stages) may be so great that they are not immediately recognized as one species. In this way a species may have five, ten, twenty, or more names (synonyms), each with its own stream of literature. To join these streams together requires careful study of the original material, often in several other museums, and a large number of specimens from perhaps as widely separated localities as Tokyo, Cape Town, and Trinidad. The taxonomist is constantly trying to decide whether the differences that he finds are due to individual variation, or whether they are indeed indications of a distinct species.

The loss of so much early material, such as that of the Sloane collection, is a pity. Far more important, however, are all those subsequent specimens that have formed the basis for new scientific names from the time that our modern system of nomenclature began (with the works of Linnaeus in the mid-eighteenth century). The specimens used in a description of a new species are referred to as the 'types', and these must be used as the final reference point for the identity of the species whenever the question of its correct name arises. In deciding whether two or more names really apply to the same species, the types of all of them must be examined and compared (as well as other material). The Museum has very large numbers of such types, partly because so many new species were described at the Museum, and partly because type specimens are sometimes deposited here by outside workers when they describe a new species.

Before the invention of the camera, naturalists either drew their material or employed artists to do so, and even now a good biological drawing is often preferred to a photograph because the artist can select the features he wants to emphasize. In some cases the original specimens were either not kept or have since disappeared, so that the drawing itself becomes the type. Early drawings may also be critical in identifying material from expeditions, perhaps to give invaluable data on the distribution of certain species before the area was colonized and the habitat altered. The numerous and often very large collections of drawings in the Museum are a frequent source for the taxonomist, one of the most important being those from the Cook voyages.

Taxonomic work is impossible without a good library. Just as early specimens must be kept for future work, so also all the early literature in which those specimens were described. The growth of the Library at the Museum has paralleled that of the collections and although rather few books were transferred from Bloomsbury when the Natural History Departments were moved to South Kensington, the Library now is one of the greatest of its kind in the world, both for modern books and journals and for the older and often very rare literature.

To attempt to identify and classify all living and past species of animals and plants, as well as all minerals, requires huge collections of both specimens and literature which never really lose their usefulness. More than any other biological discipline, taxonomy is thus 'collection-based'.

A beautifully carved *Nautilus* shell, one of the objects remaining from the museum of Sir Hans Sloane (which formed the basis of the Museum's collections). It dates from the late seventeenth century, when the carving and decoration of such shells reached a peak of accomplishment in the Netherlands, the most famous being produced by members of the Belquin or Belkien family. This particular shell is thought to have been executed by Johannes Belkien, son of the famous mother-of-pearl engraver and inlayer Jean Belquin, but little is known of him. It shows well the three main techniques used by the Dutch seventeenth-century workers: cameo work, engraving, and the cutting of the chamber walls, in this case to produce the semblance of a helmet.

This superb carnelian bowl was once the property of Sir Hans Sloane. Unfortunately, a considerable amount of Sloane's material was among the duplicate and unwanted specimens sold by auction in 1803 and again in 1816; at the same time, a quantity of material thought to have no scientific interest was thrown away, including Sloane specimens. In 1837 the modern system of registering mineral specimens was begun, but it was not until 1883 and shortly after the move from Bloomsbury to South Kensington that the earlier material was catalogued. In the 1930s, a serious attempt was made to locate the remnants of Sloane's 10 000 or so mineral specimens. Less than two hundred could be recognized, among which were a flawless emerald, a very fine sapphire, a Roman skull encrusted with travertine, and various beautifully carved cups and trinkets in varieties of silica. Nowadays, such objects are treated with the respect they deserve.

These horns, from an Indian water buffalo (*Bubalus arnee*), each measure 1.96 metres and are the largest of their kind ever recorded. They were originally part of Sir Hans Sloane's collection and were described by him in the *Philosophical Transactions* of the Royal Society in 1727. Apparently they were found by a certain Mr Doyle in a cellar in Wapping and, either because he could not pay his bill or because he wanted to make some extra gesture of appreciation, were given to Sloane in return for attending on Mr Doyle during some illness. Sloane was an excellent physician, with a flair for accurate diagnosis, but although he left an estate of over £100 000, he claimed never to have refused to see a patient who could not afford his fees. It is nice to think that Sloane's generous nature, together with the gratitude of the long-forgotten Mr Doyle, are commemorated in this magnificent pair of horns.

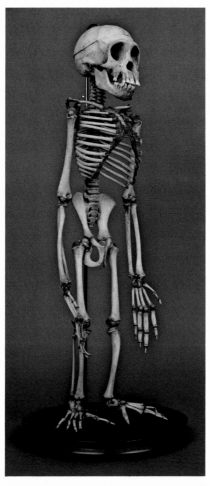

Known as 'Tyson's Pygmie', this young male chimpanzee was brought to England in the late seventeenth century but it died not long after its arrival in London, having suffered a fall on board ship which rather seriously damaged its jaw. It was then about eighteen months old. Its body was acquired by the physician Edward Tyson, who dissected it carefully and gave an account of it in his *Orang-outang, sive Homo sylvestris: or the anatomy of a pygmie*, published by the Royal Society in 1699. Tyson regarded it as a new species, nearer to man than monkeys, and probably identical with the pygmies of classical literature. His use of the name *Orang-outang* (man of the woods) probably led to Lord Monboddo's notion that an orang-outang could be turned into a gentleman if properly brought up. In turn, this inspired Thomas Peacock's satirical novel *Melincourt*, in which the amiable Sir Oran haut-on is bought a baronetcy and a seat in Parliament. Tyson's book, a worthy forerunner of Thomas Huxley's *Man's place in nature*, helped to support the view that there was a 'great chain of being', from the lowliest to man at the top.

The Museum's collections contain many specimens brought back from famous expeditions. Perhaps the most dramatic are the 12 kilograms of rock specimens collected by Captain Scott on his last expedition to Antarctica in 1912. They were found by the relief party when they discovered the bodies of Scott and his two companions in their tent. In spite of their failing strength, Scott had refused to abandon the specimens, believing them to be of considerable scientific interest. They are now part of the Museum's rock collection and a small selection is illustrated here: slate (*top left*), crushed sandstone (*top right*), pyrite (*lower left*), and biotite granite (*lower right*).

These shells, in their original metal boxes and occupying seven drawers, were once part of the collection of Sir Joseph Banks, who had sailed around the world and collected some of them during the first of Captain Cook's three historic voyages (1768–71). On his return from the voyage, Banks continued to enlarge his natural history collections, kept from 1777 at his house in Soho Square. His main interest, however, was in botany and gradually he gave his zoological specimens away, one large part going to the British Museum and another to the surgeon/anatomist John Hunter (to become part of the museum of the Royal College of Surgeons). The shells and the insects were donated to the Linnean Society in 1815, and in 1863 these were in turn given to the British Museum.

Darwin's collection of corals and other reef-building organisms from the Cocos-Keeling Atoll in the Indian Ocean was made during the cruise of the *Beagle*. Although these were the only reefs that Darwin actually visited, his well-known theory of coral-reef formation basically holds good today. From his brief experience and from the literature, he came to the very bold conclusion that the ocean floors were in fact subsiding, an idea which has now been shown to be correct. These specimens are not directly relevant to Darwin's ideas on organic evolution, but they are related to his first published exercise in the notion of 'evolution', that is to say progressive change, in this case the geological origin of coral reefs through alterations in features of the earth's crust.

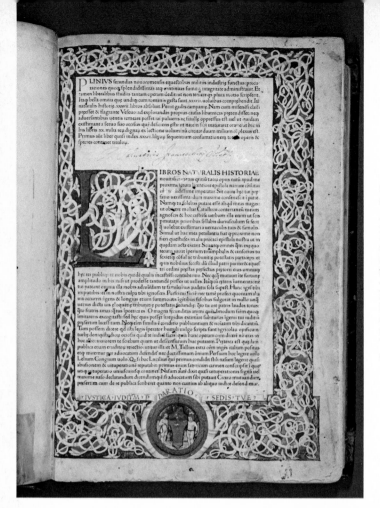

One of the oldest books in the library, this first edition of Pliny's *Natural History* was published in Venice in 1469, not long after the invention of printing in Europe. Pliny the Elder (23–79 AD), although a busy Roman official and a high-ranking officer in the army and navy, never lost a moment in the pursuit of scholarship, even in the baths dictating or being read to. The *Natural History* is his major work, written in 37 volumes and attempting to encompass all that was known on the subject, whether fact or fable. Pliny has been accused of being uncritical, but in fact he attacked magic and superstition and his final act was to insist on observing more closely the eruption of Vesuvius and the destruction of Pompeii, an act of true scientific curiosity that cost him his life.

This curious insect collection is not only one of the earliest that still survives, but is perhaps unique in the way that the specimens have been mounted between two sheets of mica. In fact, it probably owes its continuing existence to this method. It once belonged to a great figure among the early collectors, James Petiver (1663/4–1718), an apothecary who for 23 years lived unmarried at 'the sign of the White Cross in Aldersgate Street' in London. Petiver amassed a huge collection of plants and animals, making a point of befriending sea captains and giving them careful instructions on what to look for and how to preserve it. The bulk of Petiver's varied collections was bought by Sir Hans Sloane, apparently for some £4000, and these eventually came to the British Museum. Among the insects, contained in two large volumes, is a dragonfly with its exact date of capture (27 August 1700); rarely were such details recorded in those days. There are also specimens mentioned by Linnaeus in his *Systema naturae* of 1758.

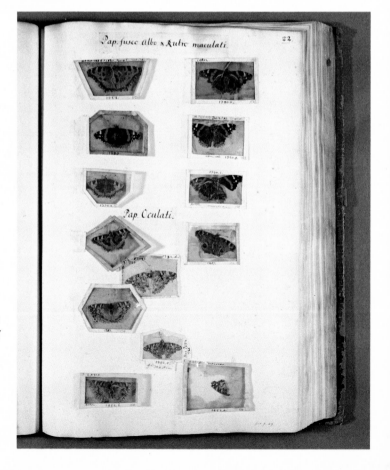

Of all the animal specimens brought back from Captain Cook's first voyage around the world, none excited more interest than the kangaroos (two or three skins and skulls). Sydney Parkinson (1745–71), who died on the return voyage, made this sketch on the eastern coast of Australia at the Endeavour River. It is almost certainly the first European drawing of one of the larger kangaroos, but unfortunately it was eclipsed by George Stubbs' famous but less accurate oil painting, made from one of the skins brought back. For twenty years the Stubbs' painting was the model for numerous engravings, until live kangaroos were brought back to England in the 1790s and Parkinson's accuracy confirmed. The Museum has three volumes of animal drawings by Parkinson and eighteen volumes of beautifully depicted plants, many still of great scientific interest.

This water-colour of *Banksia coccinea* from Australia (with an actual specimen on the left), was drawn by one of the greatest of natural history artists, Ferdinand Lucas Bauer (1760–1826). Bauer based it on his initial sketch made while on Captain Matthew Flinders' historic circumnavigation of Australia in 1801–3. During the voyage he made pencil sketches, with colour notes, of more than 1500 plants. On his return and under the supervision of the botanist Robert Brown, he made completed drawings of a selection of these in water-colours. Austrian by birth, Bauer first made his name as natural history assistant to Professor John Sibthorp on a visit to the Levant. His artistic work was evidently much admired by Sir Joseph Banks, who was instrumental in his appointment as natural history draughtsman on the Flinders' voyage. His brother Franz Andreas Bauer (1758–1840) is no less renowned as a botanical artist.

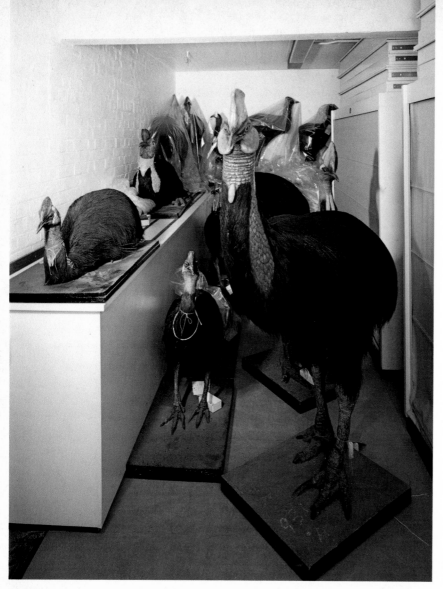

The storage of bird specimens can pose problems of space and as a rule scientific collections are made up of 'study skins' which can be stored in drawers. From such skins the external features can be studied fairly easily. For some reason Lord Rothschild decided to have no less than 65 of these large cassowaries mounted as if for public exhibition and as such they make a unique collection—and something of a headache for the curator. Cassowaries, large flightless birds of the tropical rain-forests of New Guinea, the adjacent islands, and northern Australia, seem to comprise only three species and are most closely related to the emu. Their relationship to the other flightless ratite birds—the ostrich, rheas, and kiwi—is still a matter of controversy.

The okapi, the only living relative of the giraffe, was discovered in a curious way. In 1900 Sir Harry Johnston, the British Special Commissioner for Uganda, undertook to return to their homes in the Congo forest a band of pygmies kidnapped by a German entrepreneur, who had hoped to exhibit them at the Paris Exhibition. Johnston took the opportunity also to investigate tales of a large ass-like animal known as 'okapi', but the only physical evidence he could find was some strips of hide. These formed the basis for P. L. Sclater's description of a new species of horse, *Equus johnstoni*. The following year Johnston managed to obtain a complete skin and two skulls,

which he sent to the Museum. The Director, Dr E. Ray Lankester, examined these and realized that the okapi was not a horse at all but a relative of the giraffe and he called it *Okapia johnstoni*. Shown here is a more recently preserved specimen exhibited in the Rowland Ward Pavilion of African animals; inset are the original strips of hide.

Sloane's viperfish, *Chauliodus sloani*, with its fang-like teeth, was first described and illustrated by Mark Catesby in the appendix to his *Natural history of Carolina* (completed in 1743). Apparently, the specimen shown here, the first ever examined by a naturalist, was sent to him from Gibraltar. Catesby later gave it to Sir Hans Sloane and it is the only spirit-preserved fish from Sloane's collection that has been recognized. In spite of Catesby's reputation for accurate description, the appearance of the fish was so bizarre that it seems not to have been accepted as a real fish until 1801 when it was named in honour of Hans Sloane.

With the growth of the collections the question of space and the method of preservation have constantly posed problems for the curator. The fishes shown here are preserved in jars of ethyl alcohol, a more pleasant preservative to work with than formalin but with the disadvantage that it will evaporate rapidly if the jar tops do not seal perfectly. Fish specimens were preserved as dry skins before the discovery of alcohol preservation in the 1660s, but skins or stuffed specimens were still being prepared in the nineteenth century and the Museum has many which are still of considerable scientific importance. New preservation techniques are often tried, but there is no way of knowing whether they will stand the test of time.

The botanical collections, like those of other Departments, are continually growing, with current accessions ranging from 20 000 to 30 000 plant specimens a year. These 'specimens' can represent anything from a very small part of a large individual like a tree, to a sample from a population of minute single-celled organisms which might include a hundred or more species and millions of individuals. The photograph shows the initial curation of a recent collection of over 3000 flowering plant specimens from Sicily and southern Italy. During this initial sorting, material is selected for mounting and incorporation into the Museum's collections, while duplicates are prepared for exchange with other institutions: as a result, specimens are received in return, thus further enriching the Museum's holdings.

A page from the Watling collection of early Australian drawings. Thomas Watling is famous for his simple but very honest representations of the convict settlement at Botany Bay in Australia in the 1790s, in which he also gives very precious pictures of the Aborigines and their artefacts, as well as native plants and animals. The latter are extremely useful in showing, for example, the much wider distribution of some marsupials at that time. The Library has a volume of 512 drawings, some of which are signed by Watling, but others are by fellow artists working at the same period. Watling himself came from Scotland and gave drawing lessons, but his skill led to his conviction as a forger, a crime punishable in those days by transportation. The animals in this drawing are most likely quolls (*Dasyurus viverrinus*) showing the two colour phases in this marsupial 'cat'.

The first major addition to the fossil shells in the founding collection donated by Sir Hans Sloane to the British Museum was that made by Gustavus Brander (1720–87), a Trustee of the Museum and a Director of the Bank of England. Here he is seen holding rather proudly the type specimen of *Strombus amplus*, described by Daniel Solander, the companion of Sir Joseph Banks on Captain Cook's first voyage round the world and a member of the British Museum staff from 1763. Solander wrote on the shell 'Hordel in Hampshire 1749' and his publication on the Eocene Barton Beds of Hampshire in 1766 was probably the first work on fossils in which the modern system of scientific nomenclature was used, the system devised by Solander's teacher, Carl Linnaeus. The actual shell is also shown here (inset).

For species such as the elephant shrews, whose range covers a great part of the African continent, large numbers of specimens are necessary before variation due to sex, age, season, and locality can be properly understood and allowed for in distinguishing the various species. In earlier times it was considered adequate to have only one or two reference specimens, but by the end of the nineteenth century the study of variation, stimulated by Darwin's explanation of the mechanism of evolution and by greatly increased collecting, showed that variation within a species could be considerable. Many so-called species were merely extreme forms. This need for large series of specimens led to an increasing separation between the requirements for public display and those for research.

Although diversity of one degree or another is found in all animal and plant groups, it is particularly striking in the insects. There are about 150 000 species of moths and 15 000 species of butterflies and the range of shape, colour, and pattern among them is extraordinary. The various colours and shapes obviously serve some purpose, but at times it is difficult to imagine what this might be. For example, the use of the long tails in the pink moth (*Eudaemonia argus*) is still a puzzle to zoologists. On the other hand, the horner clearwing (*Sesia apiformis*) looks extremely like a hornet wasp and is undoubtedly avoided by certain predators. The leaf moth (*Draconia rusina*) is well camouflaged in nature, its ragged leaf-life wings hardly resembling those of a normal moth. Eye-spots on *Eochroa trimena* must certainly have a startling effect if suddenly displayed, while other brightly coloured species often warn predators that they are distasteful. Moths are usually thought of as rather drab; in fact, only four butterflies are shown here (lower three in left column and one under the hornet clearwing).

Diversity

From the centre of the Earth to the surface is about 6400 kilometres, and from the surface to the extremes of the atmosphere is about another 1000 kilometres. Between these two, however, only a thin layer of about 19 kilometres is actually colonized by animals and plants, like a coat of paint on a football. The sole-like fish (or perhaps sea slug) seen by Jacques Picard and Don Walsh from their *Bathyscaphe* at a depth of 10 300 metres in the Marianas Trench of the western Pacific and the alpine choughs seen flying at nearly 8000 metres on Mount Everest during the 1953 expedition represent the approximate vertical extremes of life. The vast majority of plants and animals live within a much thinner layer, on the land surface and in the upper 150 metres of the sea—a monomolecular layer on the football, so to speak. Within this thin layer, however, are found nearly two million different kinds of minerals, plants, and animals, each with its own special characteristics and all of them of actual or potential interest to scientists in the Museum.

In earlier times, the diversity of the natural world was to some extent hidden. Microscopic life could not be seen and without the use of even a good hand-lens the differences between the smaller species could not be appreciated or suspected. For this reason, a knowledge of animals, plants, and minerals was often a very practical business, depending on the extent to which they entered social activities such as hunting, medicine, food gathering, and the search for materials to build, clothe, decorate, and so on. Nevertheless, societies still at a primitive level of culture can often supply names for large numbers of animals and plants and in many instances the species correspond rather exactly with the species as recognized by scientists; Ernst Mayr, for example, found that a tribe of Papuans in New Guinea had names for 136 of the 137 birds that he recognized in the area.

Had Aristotle and Theophrastus spent their working lives in New Guinea instead of in fifth-century BC Greece, they would surely have recorded more than the 500 animals and 500 plants to be found in their writings. It is difficult to know how many species were known in mediaeval Europe, but the big increase began in the sixteenth and seventeenth centuries when actual specimens rather than travellers' tales were brought back to Europe. In the latter part of the eigtheenth century Carl Linnaeus had bestowed names on about 4400 animals and 7100 plants. To give an idea of the subsequent growth in knowledge of the true diversity of species, one can take the most economically important single group of fishes in the Indian Ocean, the herring and anchovy-like fishes. Not one of them was known to Linnaeus in 1758, yet they were caught in perhaps thousands of tonnes along Indian coasts. By 1822 there were only 12 species recognized, in 1850 49 species, and 66 species by 1972; more recent work recognizes more than 100 species. However, this represents only two out of over 400 families of fishes, and fishes total only 1.6 per cent of all known animals. In fact, Linnaeus knew much less than 1 per cent of all the animals and plants that really exist. The figures in the table give some idea of this vast assemblage.

Approximate number of species known

Living animals		Living plants	
Sponges	10 000	Fungi	40 000
Corals, jellyfish, etc.	10 000	Seaweeds, etc.	9000
Flatworms	25 000	Lichens	18 000
Roundworms	30 000	Mosses, etc.	23 000
Molluscs	110 000	Ferns	10 000
Earthworms, etc.	15 000	Conifers, etc.	600
Spiders, scorpions, mites, etc.	130 000	Flowering plants	286 000
Crustaceans	30 000		
Insects	800 000	*Total*	386 600
Starfish, sea urchins, etc.	6000		
Fishes	20 000	*Microorganisms*	66 000
Amphibians, reptiles	8500		
Birds	8600	*Fossil animals and*	
Mammals	4000	*plants*	300 000
Total	1 207 100	*Minerals*	2500

Grand total of all known living and fossil species of animals and plants = 1 959 700 (perhaps 2 million by the end of the century).

The origin of this biological diversity has been the process of evolution operating within the context of the enormous physical diversity of the Earth's surface. Quite apart from major zones reflecting differences in climate, topography, type of surface, height or depth from sea level, and so on, there are very many thousands of different microhabitats to which perhaps only one species within a group is adapted. The filling of each of these habitats or ecological niches has taken over five hundred million years and still continues, relationships becoming ever more complex as new forms evolve ways of surviving with existing ones or take up the struggle to replace them.

Rocks are composed of minerals and it is logical that the study of the natural world should begin with minerals, of which about 2500 are known (and about 40 new ones are described each year). Popular attention usually centres on the more showy ones, either the gem stones or those that form large and spectacular crystals, but a walk through the Mineral Gallery will reawaken interest in the humble pebbles picked up on the beach last year. Minerals are usually classified according to their chemical composition and their structure; the determination of each requiring considerable skill and the use of sophisticated equipment. However, the properties of light reflected from the cut and polished surfaces of minerals or transmitted through paper-thin sections provide useful information much more readily. The Department of Mineralogy also houses collections of rocks, oceanic sediments, and meteorites. About 3000 meteorite specimens have been recovered and pieces of more than half of these are in the Museum's collections.

The study of minerals leads on to the study of plants, since soils are a crucial factor in their growth and distribution. In a sense, plants dominate the world through their ability to harness the sun's energy by photosynthesis and to maintain the oxygen-rich atmosphere necessary for animal life. Of the near 400 000 plant species, it is the *angiosperms* or flowering-plants that far outnumber the rest and show the greatest diversity. The angiosperms include the familiar grasses, flowers, vegetables, shrubs, and most trees, and produce much of mankind's needs. The other seed-bearing group is the *gymnosperms*, comprising the pines and their allies, among which are the largest and longest-living of all trees, the giant redwoods or sequoias. Most of the remaining plants reproduce by means of microscopic spores (in at least one part of their life-cycle) and among the most successful have been the fungi, ranging from the common mushroom to single-celled yeasts used in making bread, wine, and beer. The algae range from common seaweeds to the microscopic diatoms whose accumulated skeletons cover vast areas of the ocean bed, while yet another kind of plant, the ferns, are frail reminders of a group that dominated the Coal Measures forests. Most curious of all perhaps are the lichens, fungi which have formed a working relationship known as symbiosis with certain algae, and can colonize surfaces unavailable to most other plants. The richness of the plant kingdom is reflected in the Museum's collections, where of flowering plants alone there are some 1.7 million specimens.

Animal species outnumber those of plants by three to one and if this is a mark of success, then insects should be reckoned the dominant group, comprising over 60 per cent of all animals; they are followed a long way behind by the spiders, scorpions, and mites, and also the molluscs (both about 10 per cent). However, it is noticeable that the most highly evolved group, the mammals, is one of the smallest (about 4000 species or 0.3 per cent of the animal kingdom), while there are reasons to feel that just one species has dominated all, *Homo sapiens*. As in plants, the simplest forms of animals are single celled, such as the amoeba, and increasing complexity has made possible an extraordinary variety of shapes and sizes, from moths of less than 3-millimetre wingspan to blue whales of 30 metres in length or more. Unlike plants, however, by far the greatest diversity among animals, whether measured in terms of numbers of species or in the variety of fundamentally different organizations of the body, is found amongst the smaller, inconspicuous invertebrates rather than the more obvious and better known vertebrates. A spoonful of soil, for example, can contain a hundred different species, showing a range of differences much greater than those between an elephant, a mouse, a duck, and a trout.

In studying the diversity of minerals, it is not enough merely to record what they look like: a great deal can be learned from the way that they react under different conditions, and also from the associations of different minerals in rocks. This is no less true in plants and animals. It is convenient to study them as preserved specimens, but their purely anatomical diversity reflects an equal diversity in the way that they 'make a living'. The way that they are constructed (anatomy), the way that the structure operates (physiology), and for animals the way that all this behaves (ethology) are equally important aspects of diversity. In turn, these must be seen in the context of individuals within a population and again in the context of a particular species within a community of animals and plants living in a particular environment. Finally, one must remember that each of these aspects of diversity was arrived at through a long historical process—evolution. Thus the pure cataloguing of diversity, which largely occupied the earlier workers, has become an extraordinarily complex affair—and a much more interesting one.

While Hillary and Tensing were struggling to conquer Mount Everest, not far below them was the permanent home of a small dark jumping spider, *Euophrys omnisuperstes*, renowned for its ability to live at altitudes over 5500 metres in the Himalayas. The record height is held by a young specimen collected at 6700 metres on the slopes of Everest in May 1924. These spiders, which are active in bright sunshine and locate their prey by sight, are thought to be the final link in a food chain founded on wind-blown vegetation. They feed on springtails and small

sandflies and survive the freezing temperatures by resting in a silken cell spun under the rocky debris. Surprisingly, they do not show any special external adaptations to life at such high altitudes.

Colour in minerals may arise in a number of different ways. Some are naturally coloured, in that the presence of certain chemical elements imparts a colour; many manganese minerals, for example, are pink. Others are colourless in their pure state but are frequently found in coloured forms due to the presence of chemical impurities or of defects in their crystal structures; these lead to the selective absorption of certain wavelengths of light and thereby produce characteristic colours. The mineral fluorite (calcium fluoride, CaF_2) is an example of the latter group, being perfectly colourless when free from impurities or structural defects. However, the most sought-after examples are those showing deep purple, green, or yellow colours. Typical is the specimen shown here of purple fluorite from Weardale, County Durham.

In the 1840s the Manx naturalist Edward Forbes supposed that there was a lifeless or azoic zone in the oceans, beginning at a depth of about 400 metres. Twenty years later it was apparent, mainly from inspection of submarine cables, that marine organisms flourished down to some 2000 metres, and today it is known that life extends down to even the deepest of the submarine trenches such as the Mindanao Trench in the Philippines at over 11 000 metres. The deep-sea anglerfish shown here (*Melanocetus johnsoni*—10 centimetres) is found at 200 to 2000 metres. The jaws are huge and the stomach distensible in order to take advantage of meals that may not be too frequent. The 'fishing rod' on the snout is modified from an anterior dorsal fin ray and has a light organ at its tip to attract prey.

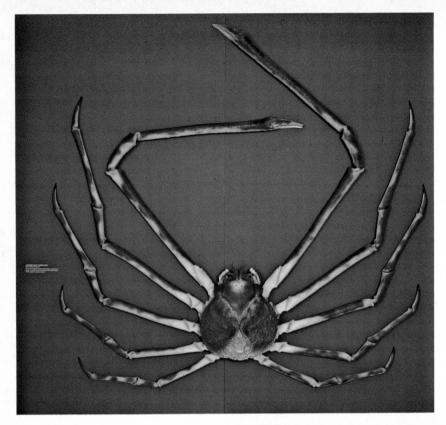

Size in the arthropods or joint-legged invertebrates (insects, spiders crustaceans) is limited by the sheer weight of a skeleton that encloses the body; after a certain size the space inside is not enough for the muscles needed for movement. This is less limiting in water and the largest of all the arthropods is a marine crab, the Japanese spider crab (*Macrocheira kaempferi*) which occurs at depths of 50–300 metres off the southeastern coasts of Japan. The body is 30 centimetres across, while the legs span up to 3 metres between the tips of the outstretched claws. Other species of spider crab are well-known for the way they attach pieces of seaweeds, hydroids, and sponges to the 'hairs' on their backs for camouflage, the old pieces being replaced by new ones when they move to a different background.

The sea-mouse (*Aphrodita aculeata*) is a common but most unusual species of polychaete worm found below the tidal zone on sandy or sand-mud bottoms. Unlike most polychaetes, the body is rather short and stocky (reaching 15–20 centimetres), in contrast to such typical elongate forms as the ragworms (*Nereis*). Equally unusual is the fine felt of hairs or chaetae on the sides, which give an iridescent sheen, and through which protrude large brown spines that can cause severe irritation if they puncture the skin. The common name derives from its curious 'hairy' look. When turned over, the typical segmented body of the true worms (Annelida, to which the polychaetes belong) can be clearly seen.

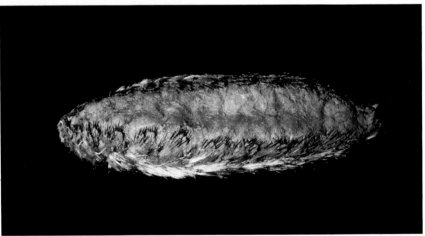

Bryozoans are aquatic colonial animals which as adults are usually attached to a substratum such as a seaweed or rock. The Australasian genus *Selenaria* is 'free-living' and *Selenaria maculata* (shown here) is able to 'walk'. Colonies are about 12 millimetres in diameter and comprise more than 2000 members, called zooids. Some zooids are modified and have elongated bristles with complex muscle systems. Their movements are co-ordinated so that the colony rotates and 'lurches' towards light, moving at about 1 metre an hour.

Less than a hundred species of sea lily exist nowadays, but in the geological past these stalked relatives of the sea urchins and starfishes were diverse and abundant. The stem is quite flexible and bears a capsule-like theca or cup which houses most of the body of the animal and bears the long feathery arms used for collecting food. The fossil shown here, *Pentacrinites fossilis*, came from Liassic rocks at Lyme Regis in Dorset. Unlike modern species, these crinoids did not attach themselves to the sea floor, at that time apparently uninhabitable, but to floating logs which would eventually sink through waterlogging or the weight of the crinoids.

This fine example of the ammonite *Stephanoceros humphriesianum* came from Inferior Oolite rocks of the Middle Jurassic near Sherborne in Dorset and dates from about 170 million years ago. Most of the shell is preserved except for a small portion through which shows the pale oolitic limestone filling the whole of the last whorl, which was the living chamber of the ammonite. Limestone also occupies the chambered inner whorls, which in life were filled with gas and provided a flotation device so that the animal could swim suspended at any depth in the sea. They resemble their nearest living relatives among the Cephalopoda, the species of *Nautilus* of the Indian Ocean and the western Pacific.

One of the most obvious aspects of diversity in the natural world is size. Among the largest and longest lived of all plants are the redwood trees of California, some of which are believed to have lived for over 3000 years. The slice shown here, which is 4.5 metres in diameter, came from the trunk of a giant sequoia (*Sequoiadendron giganteum*) felled in 1892, and from the rings it is estimated to be more than 1300 years old. The related coast redwood (*Sequoia sempervirens*) is rather taller, reaching over 100 metres in height, but its girth is considerably less.

The large freshwater amoeba commonly studied in schools (*Amoeba proteus*) can be contrasted with the testate amoebae (Testacida) which are enclosed in a single-chambered shell. In the species shown here, *Difflugia corona*, the shell is made from quartz particles cemented together by an organic substance. The shell can be spherical or oval and is about 150–200 microns long (about one fifth of a millimetre). In life, pseudopodia or 'false legs' protrude through an aperture. Reproduction is asexual, an identical daughter cell being formed by binary fission, The pseudopodia are used not only for movement but to engulf particles of food (bacteria, algae, and smaller testate amoebae). Under adverse conditions, for example when the pond or ditch dries out, the oxygen becomes short, or food scarce, these amoebae can form temporary or resistant cysts.

Of the few plants adapted to trapping insects and digesting their bodies as a supplementary food source, one striking example is the Australian pitcher plant (*Cephalotus follicularis*), shown here in a drawing by the great natural history artist Ferdinand Bauer, draftsman on Flinders' classic voyage round Australia at the beginning of the nineteenth century. Although this species is superficially similar to the more familiar Indo-Malaysian pitcher plants (*Nepenthes* species), it is actually not closely related to that group: it is restricted to southern parts of western Australia and occurs in peaty swamps. Inside the pitchers there are overhangs and downward-pointing hairs to prevent escape, below which are glands secreting digestive enzymes to assist in the breakdown of the insect's body, after which the nutritive substances are absorbed through the wall of the pitcher.

Popular interest in dinosaurs has never waned and in recent years there have been some excellent television films and books, but in the 1850s they were an even more exciting topic when the sculptor Waterhouse Hawkins made the life-sized models for the Crystal Palace (where they can still be seen today). Shown here is the skeleton of *Iguanodon atherfieldensis* (on display in the Museum), one of the most complete dinosaurs ever found in the British Isles. It was collected by R. W. Hooley in 1917 from shales of Lower Cretaceous age (115 million years ago) on the Isle of Wight. Other specimens are known from Sussex, Kent and Belgium. The species of *Iguanodon* were plant-eaters that walked on their hind feet. The model that Hawkins made for the Crystal Palace, however, was mistakenly placed on all fours and its thumb-spike was set on the end of its snout; it is so large that a famous dinner party was held inside it.

A lichen is the result of an intimate relationship between a fungus and an alga, the fungal partner being different in each of the 18 000 species of lichen. The algal partner is represented by fewer species and thus the same alga may sometimes occur in a diverse range of lichens. The alga can exist in a free-living state, whereas the fungus cannot. Lichens, as well as mosses, are generally adapted for growth on substrates not normally colonized by other plants, such as bark or rock. As a group lichens are extremely long-lived and slow-growing and in suitable habitats some may persist for up to 4000 years. Many contain unique chemicals which form the basis for their long established (but now largely outdated) use in the production of dyes. They are also employed for the dating of glacial movement and in the manufacture of some antibiotics. Their extreme sensitivity to low levels of important air pollutants, such as sulphur dioxide, hydrogen fluoride, heavy metals, and radio-active fallout, has led to their recent use as major biological monitors for the purity of the atmosphere.

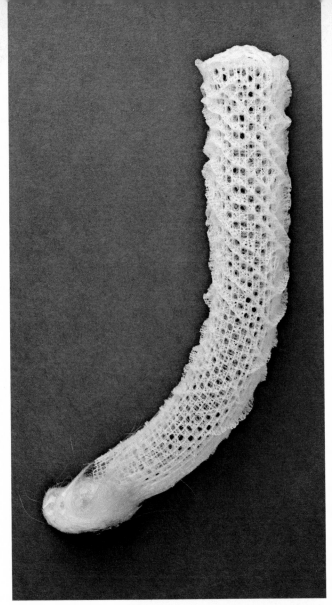

The humble bath sponge (*Spongia officinalis* and a few relatives) is a drab, inert, and lifeless object that gives little hint of the colour, diversity, and sheer ingenuity found among the 10 000 members of this group. Although the life functions operate at a cellular level, with no proof yet of a nervous system, sponges are evidently well co-ordinated, as can be seen from the beautifully symmetrical forms of some of the deeper-water species. The one shown here is the bleached skeleton of the Venus' flower basket (*Euplectella aspergillum*) belonging to the small group Hexactinellida or glass sponges (characterized by a skeleton composed of microscopic, six-armed silica spicules). This species was described by Richard Owen, first director of the Museum, in 1841. The exquisite white skeleton of these animals was much admired by the Victorians and a pair could fetch 5 guineas, to be mounted under a glass dome. They reach about 30 centimetres in length, but specimens twice as large were formerly recorded.

These hand-selected marine diatoms from Madagascar were arranged on the slide by E. Thum of Leipzig in 1880, a professional preparator who specialized in collectors' items for wealthy amateur microscopists. The preparation shows well the diversity of these single-celled plants. The frustule or box-like skeleton is composed of silica and has two valves (forming the lid and base of the box) and a girdle of several strap-like bands forming the walls of the box. Mostly, the valves are seen here detached from the frustule, but there are also several complete frustules. About 10 000 species of diatoms are known, occurring in rivers, lakes, and seas and including fossil forms as far back as the Cretaceous period (up to 136 million years ago). They have been called the 'grass of the sea' since they occur in such large numbers and are eaten by so many animals; their photosynthetic activity probably contributes 20 per cent to the world's production of plant matter, besides releasing significant quantities of oxygen.

During the voyage of HMS *Discovery* in the 1920s, two species of the nematode worm *Crassicauda* were collected from rorqual whales off South Georgia in Antarctica. Because of the length of the worms and the difficulty of dissection, only parts of these parasites could be cut away from the tissue of the whale. In one of the worms (*C. boopis*) the posterior ends were found in the penis of male whales and in the clitoris of females, while the anterior ends of the worms were located as far away as the liver, a distance of many metres, indicating that this is perhaps the largest nematode yet discovered. Shown here is the anterior end of several worms (the part in the blood system). Although many nematodes are parasitic in animals and in plants, there are large numbers of free-living forms occurring in soils and in freshwater and marine conditions—even in hot mineral springs and in beer mats!

In many species of animals and plants the range of variation between individuals is fairly small. By contrast, termites have evolved highly specialized castes with striking differences between the steriles (workers and soldiers) and the reproductives (kings and queens). Shown here is an opened queen cell of *Macrotermes* beneath a large African soil mound. Beside the rear of the queen, who lays an egg every three seconds, is the smaller king, both with eyes. Circling the queen are blind major and minor workers (with round heads and small jaws), the first gathering the food, the second tending the colony, and also a few soldiers (oval heads and large jaws) of both major and minor castes, as well as a few small white larvae. Of some 2000 named species, about 300 damage crops, trees, and building timbers.

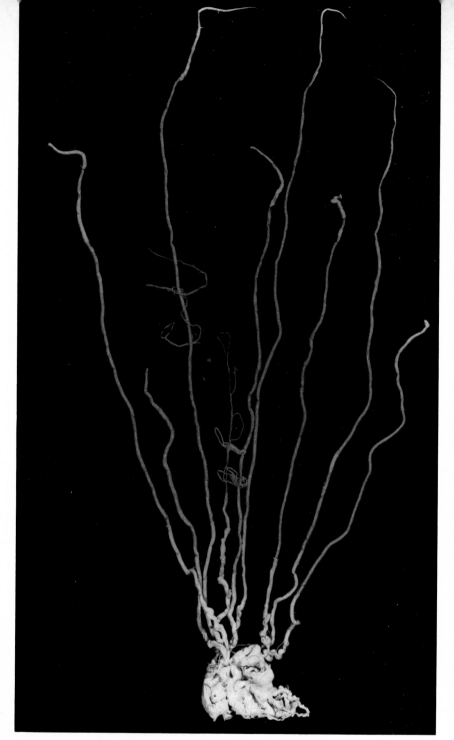

Although the demands of flight have kept birds more uniform in structure than other vertebrate groups, there is still great variation as a result of different ways of moving on land and water or getting food. Ostriches and penguins have specialized so far that they have lost the power of flight. Many wading birds have developed long legs, and in the flamingoes this is combined with filter-feeding, but unlike in ducks the head is held upside down. Dippers are quite generalized thrush-like birds, but without any striking specializations of structure they are able to walk and also to 'fly' under water. The fifth bird here is an owl, showing the end result of the trend toward forward vision in a bird that uses sight for hunting.

Man-made diversity, through artificial selection, can often rival natural diversity, at least superficially, although the ability of different strains to interbreed shows that the differences are actually not so fundamental as in wild species. This group of dogs, from the valuable collection exhibited at the Tring Museum, emphasizes the great variation that can be produced from a single species in a relatively short time. Such collections are important because they sometimes contain animals in an intermediate stage of development and very different from their present condition. Fossil evidence suggests that the domestication of dogs began some ten to fifteen thousand years ago.

In biological classification the main unit is the species. Related species are then grouped together into genera, the genera into families, the families into orders, the orders into classes, the classes into phyla, and the phyla into kingdoms. This hierarchy of groups makes the handling of more than a million species easier when information has to be summarized and it is also a means of expressing the relationships of organisms. Shown here are three varieties of the two-spot ladybird, all belonging to the species *Adalia bipunctata* (*top row*). In the next row are three members of the genus *Adalia* (*A. bipunctata*, *A. decempunctata*, and *A. deficiens*). Then come species of five different genera all within one family, the Coccinellidae (*Adalia bipunctata*, *Epilachna hirta*, *Parapriasus australasiae*, *Illeis cincta*, *Scymnodes lividigaster*). Finally, come representatives of four different families, the Coccinellidae (our original *Adalia bipunctata*), Cantharidae, Staphylinidae, and Curculionidae, all members of the order Coleoptera (beetles).

Classification

From earliest childhood we are taught to give names to the objects around us and to classify them in some simple way. As babies, we classify things for example as edible, chewable, both chewable and cuddly but not edible, warm, soft, cold, or hard. We notice size, shape, colour, texture, and ownership, but gradually through the use of language we are taught to refine these concepts and consciously to explore ever afterwards the subtle differences between objects.

The use of such classifications is normally obvious. We are concerned with the properties of an object in as much as they are significant to us, and rather than recite all those properties as a catalogue, we select just that category of immediate interest: this is the bird that lays eggs for us. The more our interest grows, however, the more we subdivide the categories, so that the bird is recognized as a domestic fowl, of a particular breed, perhaps of a certain strain or lineage, and, among the other hens in the coop, an individual with its own special personality and habits. For the biochemist or the comparative anatomist, our hen is fitted into quite different categories, perhaps to do with hormone regulation or proportions of the skull. Again, in a general way, we may also simultaneously classify our hen by means of various systems, each with their own subheadings, depending on whether we are concerned with it as a source of food, as a means of income, as a symbol of religious festival—or as a plain nuisance when it gets out into the flower-beds. The history of biological classification has been the weeding out of our special interests in the attempt to find some more universal principle.

More than any other civilization, the Greeks of classical times pioneered the route toward a classification of animals and plants based on some scientific criterion and not on limited interests (religious, culinary, medicinal, and so on, however useful these may also be). For example, Aristotle (384–322 BC) saw the animal world as a linear series, a sort of 'chain of being', but his subdivisions and analyses also reflected his need, as a busy teacher, to present a clear and simple picture to his students at the Lyceum. When later ages inherited Aristotle's works, however, they were faced with a dilemma, for here were two fundamental but opposite concepts: the idea of the box-like compartments of a hierarchy, as opposed to what is equally obvious in nature, the similarity of one kind of plant or animal to the next in almost unbroken series. The hierarchy rescued the natural world from chaotic profusion, thus equating natural history with inventory. The continuum, on the other hand, hinted at a deeper meaning and one that was simultaneously sought by religion, philosophy, and science. Both of Aristotle's ideas appeared to be true, but the explanation had to wait until the notion of evolution, and its mechanism natural selection, had taken firm root in the nineteenth century. Before the causes of diversity were known, only the results could be recorded.

Meanwhile, the rise in popularity of natural history cabinets in the sixteenth and seventeenth centuries, and the interest in natural productions that this reflected, brought forth a critical need for better classifications; the cabinets also provided the physical means to undertake such studies. To the English naturalist John Ray (1628–1705) we owe one of the bravest attempts to bring a logical classification to the thousands of animals and plants discovered since Aristotle's time. Of great importance was Ray's use of the term *species* in its modern biological sense, meaning a group of individuals that breed true and do not normally cross with other species. This is the basic unit in biological classification. With the definition of the term *genus* as a biological unit of a higher rank than species, largely through the work of the French botanist Joseph Pitton de Tournefort (1656–1708), the stage was set for the greatest of all biological classifiers, the Swede Carl Linnaeus (1707–78). From 1735 in the ever-expanding editions of his monumental *Systema naturae*, Linnaeus set out to classify everything in the three realms of nature—minerals, plants, and animals. His framework, enlarged but still used today, had seven levels or categories, descending from kingdom, phylum, class, order, and family to genus and finally species (sometimes divided into varieties). To specify the seven was exactly to locate any animal, plant, or mineral and, equally useful, to make a very exact statement about its affinities. Evolutionary ideas were already stirring, but for the time being the *Systema* was the naturalists' vade-mecum in an age when the extraordinary diversity of the natural world seemed overwhelming.

Linnaeus made yet another great contribution. He had a tidy, methodical mind and the size of his task demanded economies, not least in the naming of natural objects. Common names were imprecise and varied from one language to another, while the descriptive Latin names used by naturalists tended to become far too long and inconvenient. Linnaeus accordingly introduced a new two-name system in 1753. For learned use, the red maple could be *Acer foliis quinque lobis subdentatis subtus glaucis, pedunculis simplicisimis aggregatis*, but for everyday use he named it *Acer rubrum*—the species *rubrum* of the genus *Acer*. The advantages of this two-word or *binomial* system were soon appreciated and it has been used ever since for naming plants and animals. Because he first applied this system consistently, the works of Linnaeus are taken as the base-line for all subsequent naming (from 1753 for plants, from 1758 for animals). One result is that, to avoid cluttering the literature with duplicate names, the specialist must search back through two centuries of books and journals to be quite certain that what appears to be a species new to science has not been described before. If two or more names are found to apply to the same species, then the first validly published name takes precedence. It is for this reason that the older books and journals in the Museum's libraries are not there just for antiquarian curiosity.

Linnaeus represents the flowering of a static hierarchical classification. The next stage was the discovery that this order in nature was neither divine nor human tidiness, but the result of an underlying and dynamic process—evolution. As a result of Darwin's work, although classifications might sometimes look the same, their basis was fundamentally changed. Hierarchies, now that fossils could be added, became great branching trees rooted in an immense geological time; species had a history; and the statement that two animals or two plants resembled each other was now an invitation to discover how this had happened.

At last it all seemed to make sense, but a century later the task is still formidable. The final stage has no name, but it brings to classification a host of other disciplines and new techniques: from biochemistry and genetics to neurophysiology on the one hand, and from ocillographs and computers to the electron microscope on the other. Exciting new theories have suggested more fruitful ways to explore this 'past contained in the present', which is the diversity of natural life.

There is, however, probably no fundamental and revolutionary discovery around the corner, at any rate nothing that will materially lighten the task of providing a final classification for more than two million animals, plants, and minerals. As Darwin used to say, 'It's dogged as does it', and there is indeed a certain doggedness required of the new recruit to the Museum, who in his career may painstakingly examine and measure tens of thousands of fleas, flies, fishes, or fossils in order to update the classification handed down to him. No museum can tackle everything and the gaps must be filled in other institutions or handed down to the next generation. Nevertheless, the record of the Museum is extraordinarily impressive, with classic works in virtually every major group.

Finally, one must ask the question: Why classify? Is it only to satisfy some intellectual curiosity? To some extent, perhaps, but a much more immediate reason is that for any biological work to be meaningful, at least the identity of the material must be known and the correct name given. If the wrong name is used, or if the species is mixed or muddled with another, then the biological work can be quite useless. If the affinities of the species can be determined, then a great deal more information may be available. Thus, two closely related species not only look fairly similar but often (although not always) share similarities in physiology and habits; at least one can form a working hypothesis about how a related species will grow or behave.

A natural history museum is a 'classifying house' and although the rewards of that classifying are usually reaped elsewhere, perhaps in the more spectacular disciplines, there is nonetheless a satisfaction in providing the indispensable basis for all biological work.

Earlier workers described an enormous number of species of plants and animals, often on rather superficial characters and without probing too deeply into the anatomy of their specimens. Frequently they had only a few specimens, perhaps only one and that might be required for display purposes. Looking at the four striped grass mice (*Lemniscomys*) it would be easy to conclude that there were three species present but careful analysis of measurements made on many individuals shows that while the two at the top (*L. griselda* and *L. barbarus*) are correctly distinguished, the two at the bottom are also quite distinct species (*L. macculus* and *L. striatus*); although they live together in Uganda, there is no interbreeding between them (the hallmark, so to speak, of a distinct species).

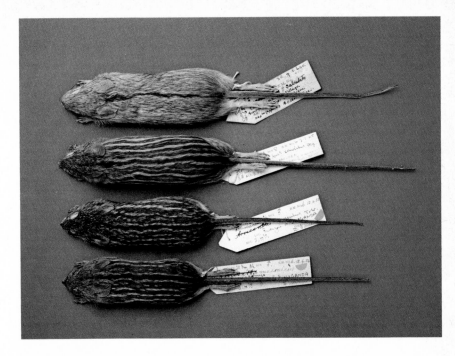

This superb water-colour of a duck-billed platypus (*Ornithorhynchus anatinus*) was made by Ferdinand Bauer (1760–1826), the natural history artist on Flinders' voyage to Australia in 1801–3. The animal had only recently been discovered in November 1797, and news of it was sent by the Governor John Hunter to David Collins, the former judge-advocate of the colony at Port Jackson in New South Wales. Meanwhile, a skin arrived in London and was examined, with considerable scepticism, by George Shaw at the British Museum. Suspecting a trick, he tried to prise the duck's bill off the skull with a pair of scissors and the marks can still be seen on the specimen in the Museum. In 1884 W. H. Caldwell showed that the platypus actually lays eggs, thus allying it with the spiny ant-eaters.

Carl Linnaeus, the eighteenth-century Swede who revolutionized the system of classification of plants and animals and who provided the basis for our modern method of scientific nomenclature, is shown in Lapland dress after his journey to the northern parts of Sweden in 1732. This mezzotint is now in the Linnean Society of London, one of the many societies founded in his honour. The seventeenth and eighteenth centuries were for natural history ones of exploration and the bringing back of thousands of specimens from all parts of the world. Linnaeus, who attempted to name, describe, and pigeon-hole all the known species of plants and animals, provided exactly the guide needed for the travelling naturalist or the bewildered museum worker faced with this wealth of new material. His classification has its shortcomings, but his method of naming species and most of his names are still in use.

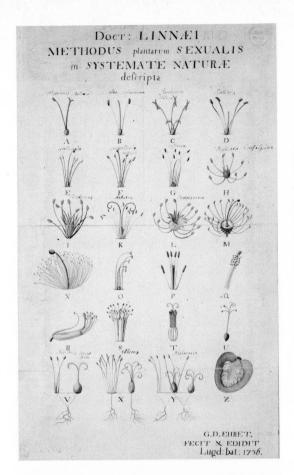

In 1735 Carl Linnaeus published the first edition of his *Systema naturae* or compendium and classification of the natural world and in it he included his 'sexual system' for the classification of plants, by which the flowering plants were divided into classes and orders according to the number of the male organs (stamens) and the female organs (pistils); the Cryptogamia were considered plants without flowers. The scheme was illustrated by perhaps the greatest of the early eighteenth-century botanical artists, Georg Dionysius Ehret (1708–70), a German who later settled in England, and from the water-colour shown here he published a number of prints in 1736 (but few survive). Ehret championed the Linnaean method of classification in England, sometimes against strong opposition, but over the next 80 years it gradually became the most widely used system in botanical works, until superseded by the work of the great nineteenth-century botanists.

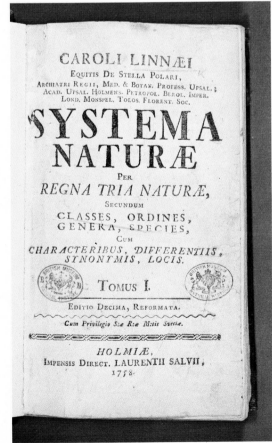

Carl Linnaeus was the founder of our modern method of giving plants and animals a binomial or two-word scientific name, the first word signifying the genus and the second word distinguishing the species (e.g. *Musca domestica*, the housefly). Since no species of animal or plant should have two or more different names, the correct name is as a rule taken to be the name that was first proposed. For convenience, the works of Linnaeus are the internationally agreed base-line for this. For zoology, the base-line is the tenth edition of Linnaeus's *Systema naturae* of 1758, whose title page is shown here. It was the enlarged twelfth edition of the *Systema* that was taken by the naturalists on Captain Cook's voyages, enabling them to find the approximate place in the system for the many new species they discovered. The starting point for botanical nomenclature is Linnaeus's *Species plantarum*, published in 1753, although there are some more recent base-lines for fossil plants, fungi, mosses, and a few other groups. The rules for zoological and botanical nomenclature are as complex as any legal system, but this is necessary to ensure the correct application of a name to a particular organism.

Similar life-styles can sometimes lead to extremely similar outward appearances, often because there is only one wholly efficient way to solve a particular problem. In this case, streamlining of the body and shaping of the fins are almost identical in a reptile (a fossil ichthyosaur—*upper*) and a modern mammal (white-sided dolphin—*lower*). The ichthyosaur is a young specimen of *Ichthyosaurus acultirostris* from the Lower Jurassic of West Germany (about 185 million years ago) and it is so well preserved that the impression of the skin can be seen, including the tail and dorsal fins and the hydrofoil shape of the flippers. Ichthyosaurs, which appear to have borne live young, were powerful swimmers of the open seas and were one of the most successful of the aquatic reptile groups until the later part of the Cretaceous period.

The amphibians include three highly specialized groups: the frogs and toads, the newts and salamanders, and the burrowing and worm-like caecilians of the tropics. Their ancestors, the labyrinthodonts of the Carboniferous to the Trias (325–200 million years ago), sometimes bore a superficial resemblance to the crocodiles, as the photograph shows. This is a reconstruction from the only known specimen of *Paracyclotosaurus davidi*, whose skeleton was 2.75 metres long. The original is not the skeleton itself but a natural mould inside an ironstone nodule, from which the individual bones were cast in plaster and then assembled to produce an extraordinarily complete specimen.

A serious problem in trying to trace the course of evolution is to decide whether similar features really indicate close relationships, or whether they have been arrived at independently by what is called convergence. In this case a thylacine (*Thylacinus cynocephalus*) and a banded duiker (*Cephalophus zebra*) have remarkably similar colour patterns. Both are mammals, but they are in no way closely related since the thylacine is a carnivorous marsupial from Tasmania, while the duiker is a herbivorous ungulate from West Africa. In this case there are many other features, both external and internal, that clearly show the animals to be fundamentally different and thus to have arrived purely by convergence at a similar colour pattern, but other cases of convergence are not so easy to detect.

Among the many pitfalls confronting the taxonomist when trying to sort out the species in a group is the sometimes considerable difference between males and females, in shape, size, colour, or other features. One of the most striking examples is the eclectus parrot (*Eclectus roratus*) of the East Indies and northeastern Australia. The male (*right*) is mainly green with red on the sides of the body and the upper part of the beak coral; the female (*left*) is totally different, with the head and body generally red, but dull purple across the lower breast and the beak wholly black. One can hardly blame the early ornithologists, who until 1874 called the males from New Guinea *Eclectus polychlorus* and the females *E. linnaei*. The colourful parrotfishes of tropical waters have suffered the same confusion, compounded because the colour also changes with age.

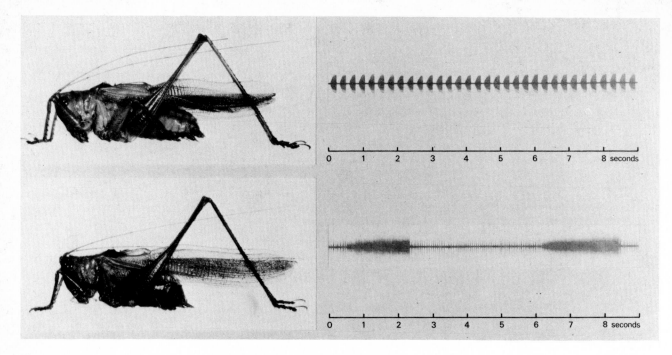

The sounds made by insects such as crickets and grasshoppers are highly characteristic and nowadays can be used to distinguish between very similar species. The two common African bush-crickets shown here (*Thyridorhoptrum senegalense*, *above*, and *T. baileyi*, *below*) are almost indistinguishable and were thought for the past century to be the same. Recent research in the Museum, stemming from the discovery of strikingly different 'songs' in the males (as shown by the oscillograms here), led to the discovery of subtle differences in the structure of the sound-producing apparatus on the forewings and has firmly established that there are two distinct species of similar habits and distribution. As found in other cases, the species are probably prevented from interbreeding because the females respond only to songs from their 'own' males. The study of insect songs is one of the many methods now used in the Museum to investigate pairs or larger groups of superficially similar species.

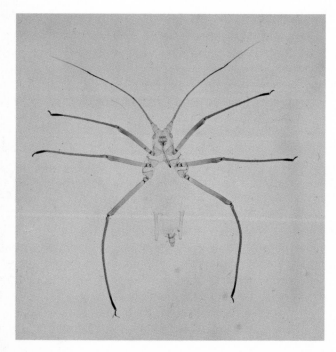

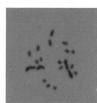

Since each species has its own unique characteristics, it is of great importance to be able to distinguish between species and to be certain that this distinction is correct. Large green aphids of the genus *Amphorophora* occur on raspberries and also on blackberries and externally the two look virtually identical. However, those on blackberries have 20 chromosomes (the thread-like bodies in the nucleus that carry the genetic information), while those on raspberries have 18. From this it has been possible to demonstrate that these are indeed distinct species, the one a raspberry pest that transmits important virus diseases, the other a common aphid on brambles of little economic importance. In the photograph is a slide-mounted specimen of the raspberry aphid (*Amphorophora idaei*) and a comparison of its chromosomes (*above*) and those of the blackberry aphid (*A. rubi*, *below*).

Aphids occur in a number of different forms or *morphs*, so that individuals within one species can differ greatly in appearance. For example, colonies of wingless individuals may give rise to winged *morphs* when overcrowded. The photograph shows a wingless adult female (*right*) of the red form of *Macrosiphum amygdaloides* (an aphid which lives on wood spurge); on the left is a winged adult female of the same species and two immature individuals (nymphs) at different stages of development, the one at the top being newly born. In the winged aphid the thorax is dark and hard to form a rigid box to house the flight muscles. In the wingless female the thorax and the abdomen form a large and thin-walled bag containing a large number of developing embryos. The nymphs have a light covering of blue-grey wax dust. These are all parthenogenetic *morphs*, able to reproduce without males, but in the autumn males and sexual egg-laying females (which look different again) are produced.

In the water buttercups (genus *Ranunculus*, subgenus *Batrachium*) there is a very striking difference in the shape of the leaves above and below the water. The submerged leaves are made up of branching filaments, whereas the floating leaves are broad and merely have the edges indented. Modern work has shown that this is largely a direct response to the environment. In other instances, in both animals and plants, a superficial difference in form between members of the same species living in different environments can be confusing for the taxonomist, who may be misled into thinking that there are two species.

51

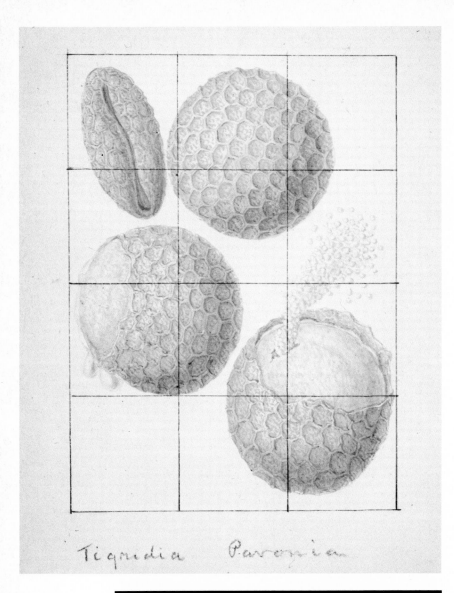

Tigridia Pavonia

It is often assumed that detailed studies of the minute parts of animals and plants did not begin until the use of modern research microscopes. This is far from true, as this illustration of pollen grains shows. They were drawn by Franz Andreas Bauer (1758–1840), brother of Ferdinand (who had sailed with Robert Brown on Flinders' voyage around Australia in 1801–3). The two Bauers were perhaps the most talented of all the natural history artists of their time. Franz Bauer was one of the first to realize that the pollen grains of different plants often look quite dissimilar. He observed and drew the pollen of many species, using the comparatively unsophisticated microscopes of the time, but he achieved a remarkably high standard of accuracy. Shown here is the pollen of the tiger flower (*Tigridia pavonia*).

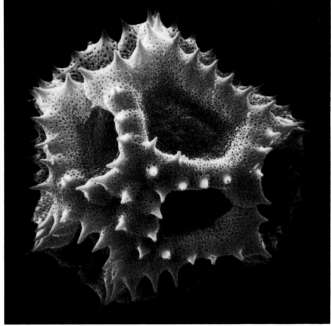

For the early naturalists, pollen was merely a dust, but the microscopists of the late seventeenth century and men like Franz Bauer at the end of the next century showed that pollen grains have a quite surprising diversity in shape and ornamentation. We now know that many plants have their own characteristic pollen and that plants with similar pollen grains are likely to be related to one another. Thus, pollen is one more character that can be used in deciding the relationships of plants. The scanning electron microscope is one of the best tools we have for studying pollen diversity. Shown here is a pollen grain from the spiny sow thistle (*Sonchus asper*).

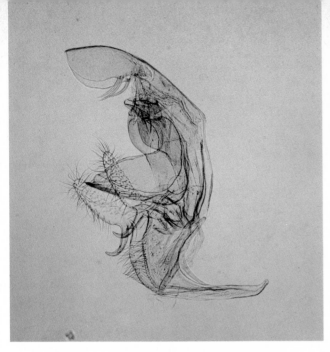

Some species of small moths cannot be distinguished from their relatives externally, but internal characters can provide the answer. In the males the genitalia are equipped with highly complex clasping organs which hold the female during copulation. Each part has a specific purpose and unless it functions correctly the right response by the female will not be triggered and the male will be rejected. This provides an important barrier to interbreeding. After dissection of the genitalia, which requires great skill, the tissues are often stained and mounted on glass slides. Shown here are the genitalia of two common British British species, *Bryotropha desertella* (*left*) and *B. terrella* (*right*), only reliably separated by their quite dissimilar genitalia.

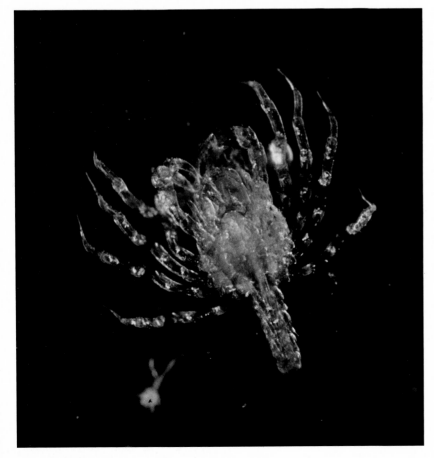

Clues to the identification and particularly the classification of animals can sometimes be found by studying the early life stages, especially when these are very different from the adult. Shown here is a small crab megalope, being the final larval stage of the edible crab (*Cancer pagurus*). Museum workers, who in the past largely relied on preserved adult material, are now rearing such species in the laboratory. From this kind of work have come useful clues to the relationships of various species and to the evolutionary history of particular groups. Such studies can be called developmental taxonomy. They also assist biologists in the important task of identifying larvae in plankton hauls at sea, thus providing better documented material for the museum worker.

Darwin was the most diffident of revolutionaries, yet Darwinism carried all before it and is still the central theme in evolutionary studies. On 24 November 1859 his *Origin of species* was published, which at last provided an overall pattern for understanding organic nature. To the furor that his book caused, because of its contradiction with the literal sense of Genesis, was added further fuel by his *Descent of man* of 1871, in which man's animal ancestry was fully spelled out. This fine statue of Darwin, now placed with that of his famous champion Thomas Huxley (1825–95), was unveiled by Huxley on the staircase of the main hall of the Museum on 9 June 1885. It is nice to know that Darwin's untidy wife Emma was known as 'little Miss Slip-slop' and that few Victorian fathers could have earned such endearing tributes to their gentleness and patience as he did from his children.

Evolution

Although Charles Darwin (1809–82) was primarily a naturalist, with little interest in philosophy, theology, or social issues, he joins Karl Marx and Sigmund Freud in dropping one of the three great intellectual bombshells of the nineteenth century. Things were never quite the same again. 'No work of our time has been so general in its influence' wrote a contemporary when discussing Darwin's masterpiece *On the origin of species*, published in 1859. Exactly how and why Darwin's evolutionary theories made such an impact has been discussed and argued about ever since, but there is no doubt that Darwinism still provides the basis for biology even a century later.

Had Darwin lived in Sicily around 450 BC, he would have nodded some approval at Empedocles, who rejected the prevalent idea that a creator had designed the world and its inhabitants with some end in view. For Empedocles, the origin of species came through chance amalgamations of their various parts and organs, which until then had wandered around independently. This may seem absurd—and Aristotle seems to have laughed uproariously at the idea of legs and arms and eyes floating around—but Empedocles made two important contributions which had to wait until the nineteenth century before they were properly developed. He turned the origin of living creatures from a mythical to a natural process and he hinted at natural selection—Darwin's mechanism for evolution—by supposing that only the appropriate combinations of body parts actually survived, the badly adapted perishing.

For the next two thousand years, almost as an accident of history, the biblical account of the Creation distorted attempts to understand the origin of life on our planet. Right up until after the time of Darwin the literal interpretation of Genesis was all too frequently used to discourage attempts at a more natural explanation, while the story of Noah and the Flood limited any ideas on the antiquity of fossils. It was indeed fossils that first prompted thoughtful men to consider the age and development of the Earth and once again it was certain Greek scholars, such as Pythagoras, who correctly concluded that fossils were the remains of once living animals and plants. This might seem obvious, but even at the beginning of the eighteenth century there were many who believed that fossils were mere tricks of nature, arbitrary condensations of vapours in the rocks, or mistakes into which the Creator had not troubled to breathe life. Boccaccio (better knwon for his *Decameron* of 1358) and Leonardo da Vinci in the next century, however, both concluded that fossils had been living creatures, while the potter Bernard Palissy in his *Discours admirables* of 1580 made the important statement that some species of fossil shells that he found were 'of a sort which is unknown to us' and only existed as fossils.

Here, one would think, was the clue, but the unchanging nature of species and their creation by a divine act as outlined in Genesis allowed only one conclusion: fossil species were those that had succumbed in the Flood (whose month was even argued from the state of fossil fruits). The date of the Creation had been fixed by Bishop Usher in the 1650s as exactly 4004 BC and it was within this narrow framework that the 'Deluvialists' had to work. However, the wealth of fossils and the variety of rocks from which they came, as well as the discovery (largely by William Smith) that particular fossils were found in particular strata, all tended to discredit this limited view. A new school, the 'Catastrophists', of which the great anatomist Georges Cuvier (1769–1832) played the principal role, suggested a series of floods, of which Noah's was the most recent, thus allowing a form of divine evolution but holding firmly to the belief that each series of creatures created after the various floods contained unchangeable species. Charles Lyell (1797–1875), in his *Principles of geology* of 1830–3, showed the absurdity of these alternate creations and destructions, as had another Scottish geologist, James Hutton, some 50 years earlier, but the fixity of species was still generally accepted by naturalists.

In a sense, the stage was now set for Darwin. Like all great innovators, he had his precursors, but likewise he produced a theory that surmounted all or most of the criticism that had been levelled at those who came before him. In a philosophical way, Montesquieu, Maupertius, and Diderot in the late eighteenth century had believed that species change, while Darwin's own grandfather had accepted the transmutation of species in his book *Zoonomia* of 1794. As in all scientific theories, however, the real point was to show *how* the theory worked, since anybody can propose a new theory. Jean Baptiste Lamarck (1744–1829) proposed an attractive evolutionary theory, based on the inheritance of acquired characters, but it lacked a plausible mechanism.

The crucial turning point in Darwin's ideas was his celebrated voyage as naturalist on HMS *Beagle* in 1831–6, a five-year voyage that took him round the world. On his return, he was convinced that species could change and he recorded in a notebook that what he had seen of South American fossils and the animals on the Galapagos Islands (especially the various finches) had provided the 'origin of all my views'. For the next twenty years he patiently accumulated his facts, filling notebooks with data, sketching out a theory of evolution, and trying to come to terms with the enormous step he was taking. The climate of opinion at the time can be judged from his remark to a friend: it is like confessing a murder, he wrote. He hesitated to publish, but, urged by Lyell, he began writing and was about halfway through when he received the famous letter from Alfred Russel Wallace (1823–1913) enclosing a manuscript embodying Darwin's whole theory. In a few hours, during a bout of malaria on the island of Ternate in Indonesia, Wallace had suddenly grasped the mechanism of evolution. The happy outcome was a joint paper read to the Linnean Society in London on 1 July 1858. The following year came publication at last of Darwin's *Origin*.

Darwin's strategy is perhaps deceptively simple. Individuals of a species show considerable variation in nature, some being larger than others, or stronger, or differently coloured, and so on. Looking at domestic animals, one can see that as much or more variation has been induced by *artificial* selection by mankind. Therefore, is there a process of *natural* selection that could, over the ages, gradually change a particular species until its descendants look as different as a daschund from a wolf? Darwin, as also had Wallace, found the answer in Malthus's *Essay* on population and he transferred the struggle for existence in human populations to the vastly fiercer world of nature. In this way, Darwin took the evidence of evolution, which had been misinterpreted or generally disbelieved, and showed that the process of natural selection was quite sufficient to have brought it about.

The storm that broke, once it was realized that man was also part of this evolutionary process, is now part of history. In 1900, with the rediscovery of the work of the Augustinian monk Gregor Mendel (1822–84) on his breeding experiments with varieties of garden peas at Brno (published in 1865 but overlooked), genetics or the mechanism of inheritance entered Darwinism. Biologists have ever since been refining the general principles that Darwin propounded in the light of modern genetic theories.

The impact of Darwinism on work at the Museum has been profound, but the initial controversies that it generated seem to have been more in the personal beliefs of the scientists than in the way that they carried out their work. Richard Owen was strongly opposed to it and was a follower of Cuvier's successive catastrophies; his successor as Director at South Kensington, William Flower, was in favour; the two outstanding zoologists at the Museum, J. E. Gray and Albert Günther were against it. It is almost as if the presence of so many neatly shelved and labelled specimens cried out for the fixity of species. In any case, a great deal of the work at the Museum at that time was a desperate attempt to catalogue what was known.

In modern times, the relationship between evolutionary theory and the classification of animals and plants is a mainspring for taxonomy. At the simplest level, it guards against the overhasty placing of two species or groups of species in the same category simply because they look alike or share some similar feature. The similarity may have been arrived at quite independently, or it may be a primitive feature shared by many other species and thus no clue to the exact evolutionary relationships and affinities of the plants or animals concerned. At a deeper level, patient breeding experiments or observations on the early life histories or behaviour patterns of animals may be needed before their true relationships can be discovered. It is possible to reject or ignore evolutionary theory and still be a taxonomist—but not a very good one.

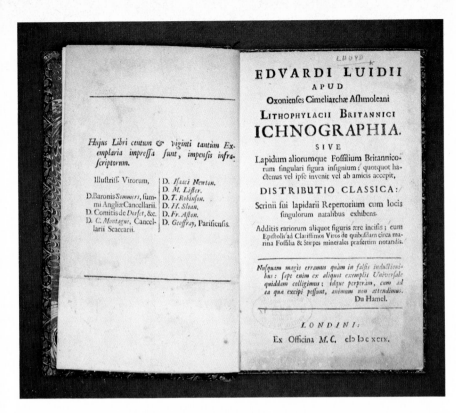

EDVARDI LUIDII
APUD
Oxonienfes Cimeliarchæ Afhmoleani
LITHOPHYLACII BRITANNICI
ICHNOGRAPHIA.
SIVE
Lapidum aliorumque Foffilium Britannico-
rum fingulari figura infignium; quotquot ha-
ctenus vel ipfe invenit vel ab amicis accepit,

DISTRIBUTIO CLASSICA:

Scrinii fui lapidarii Repertorium cum locis
fingulorum natalibus exhibens.

Additis rariorum aliquot figuris ære incifis; cum
Epiftolis ad Clariffimos Viros de quibufdam circa ma-
rina Foffilia & Stirpes minerales præfertim notandis.

Nufquam magis erramus quàm in falfis inductioni-
bus: fæpe enim ex aliquot exemplis Univerfale
quiddam colligimus; idque perperàm, cum ad
ea quæ excipi poffunt, animum non attendimus.
Du Hamel.

LONDINI:
Ex Officina M. C.   cIɔ Iɔc xcix.

Hujus Libri centum & viginti tantùm Ex-
emplaria impreffa funt, impenfis infra-
fcriptorum.

Illuftriff. Virorum,

D. Baronis Sommers, fum-
mi Angliæ Cancellarii.
D. Comitis de Dorfet, &c.
D. C. Montague, Cancel-
larii Scaccarii.

D. Ifaaci Newton.
D. M. Lifter.
D. T. Robinfon.
D. H. Sloan.
D. Fr. Afton.
D. Geoffray, Parifienfis.

From early times, fossils of shells and other objects were known, but it was many centuries before they were generally recognized as the remains of once-living animals and plants. Edward Llwyd (1660–1709), author of the *Lithophylacii Britannici ichnographia* shown here, was among the last of the scholars who believed that fossils were derived from moist seed-bearing vapours which had risen from the seas and penetrated the rocks. Two centuries before, Leonardo da Vinci (1452–1519) had understood the true nature of fossils, but biblical teaching and stories of the Flood had confused the issue. To Llwyd, however, who became Keeper of the Ashmolean Museum in Oxford in 1691, must be given the credit of publishing the first major work describing and illustrating British fossils, the *Lithophylacii* of 1699 being a summary of his own large collection.

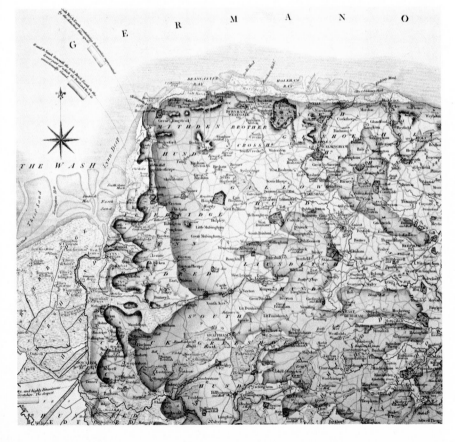

William Smith (1769–1839) has been dubbed the 'Father of English Geology' as a result of his discovery that each stratum of rock could be recognized by the fossils it contained and that the same succession of strata could be observed in different places. In this way the science of stratigraphy was born, leading to a proper assessment of the age of rocks and the gradual evolution of life forms. Smith had little formal education, but as a self-taught surveyor and later Superintendent for the Somerset Coal Canal he carefully observed the rock strata and in 1799 he drew up a *Table of strata*, from the Coal Measures to the Chalk. In 1815, after many difficulties, he published the earliest large-scale geological map of England and Wales. Shown here is part of one of his 21 county maps issued between 1819 and 1824.

Few fossils are so celebrated as that of the 'bird-reptile' *Archaeopteryx lithographica*, dating from 147 million years ago. It is not generally realized that only five such fossils exist, not all of them complete. The one shown here was the first to be discovered. It came from a quarry of Upper Jurassic lithographic limestone in southern Germany and was purchased by the Museum in 1862. It still retains many reptilian features, such as the teeth in the jaws, the long bony tail and the claws on the forelimbs. However, the feathers, whose fine structure is beautifully preserved in this specimen, are indistinguishable from a modern bird's plumage. It is probably the most perfect intermediate or 'missing link' between two classes of animals.

The early history of the Earth, before the evolution of living matter, is no less interesting to natural history than the study of fossils. Meteorites have not suffered the continuous and dynamic activity which has reshaped the surface of the Earth since it was formed and they can thus provide information on the development of the solar system. Shown here is a thin section cut from a stony meteorite that fell at Barwell in Leicestershire in 1965; this specimen is about 4600 million years old. The rounded particles, known as *chondrules*, are set in a groundmass of mineral fragments. Some 500 meteorites reach the Earth's surface every year, of which only five or ten are actually seen to fall. Altogether, over 3000 different meteorites have been recovered and pieces of more than half of these are in the Museum's collections.

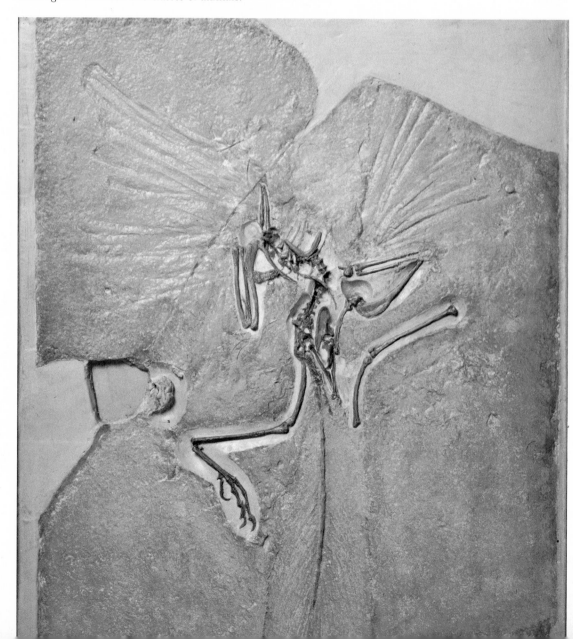

This block of Old Red Sandstone from the Lower Devonian period (about 400 million years ago) contains the skeletons of a dozen cephalaspid fishes. These strange armoured animals were once thought to be trilobites, early members of the group that contains the joint-legged animals (arthropods, such as crabs, spiders, and insects). Later they were thought to be early catfishes or sturgeons, but they are now recognized as jawless fishes related to the lampreys. Since the head is internally well-ossified, the cavities for the organs within it being lined with bone, it has been possible to reconstruct the brain, nerves, blood vessels, and other features that otherwise could only be guessed at. The group of fishes shown here was found in the railway cutting at Ledbury in Herefordshire and is a good example of the contributions made to palaeontology by the nineteenth-century development of the railways.

These bizarre Devonian fishes were discovered in 1831 by the Scottish geologist Hugh Miller (1802–56), who at first thought them to be a link between a tortoise and a fish. As a creationist, the notion of such an intermediate must have caused him some anxiety, but fortunately Louis Agassiz, the eminent natural historian and glaciologist, considered them to be a kind of armoured fish, albeit of a most unusual type. To try to understand the fossils, Miller built this paper model—about 20 centimetres long—shown here with an actual specimen of *Pterichthyodes*, its pectoral appendages outstretched in cruciform fashion. These fishes are found in lake and river deposits of the Old Red Sandstone, dating from some 370 million years ago. The paper model was presented to the Museum by Professor D'Arcy Thompson in 1898.

While Dr Gideon Mantell, a medical practioner and amateur palaeontologist living at Lewes in Sussex, was attending a patient in the spring of 1822, his wife wandered along the road and spotted some curious fossil teeth embedded in stones intended for road-mending. Her husband immediately recognized them as belonging to some hitherto unknown herbivorous animal and published an illustration of them in 1822. Unfortunately, the great palaeontologist and anatomist Georges Cuvier in Paris considered them to be from a rhinoceros and the bones which Mantell subsequently found in the quarry from which the stone had come to be from a hippopotamus. Mantell persisted and in 1825 published an account of the new reptile *Iguanodon*. He later sold the teeth to the British Museum.

This curious fossil animal, *Cothurnocystis elizae*, seems to bridge the gap between the echinoderms (sea urchins, starfishes, and their relatives) and the early vertebrate animals. It has an echinoderm-like skeleton combined with gill-slits and a distinct tail. It suggests that the echinoderms and the chordates are related to each other and share a common ancestor. It was first described by Francis Bather of this Museum in 1913. The generic name comes from its boot-like appearance (*cothurnos*, Greek for boot); he named the species after its collector, Mrs Eliza Gray. The remarkably complete state of preservation shows that it died through rapid burial in sediment.

This impression from the wing of an extinct dragonfly-like insect (*Erasipteron bolsoveri*) was found by chance by Malcolm Spencer, a miner, 270 metres below ground in the Bolsover (Derbyshire) coal mine in 1978. From this wing it can be estimated that the insect had a wingspan of around 20 centimetres, which is somewhat larger than in any living species. From the same mine an even larger wing was found, which came from an insect with a wingspan of 50–60 centimetres. This shows that at least two large species were flying through the Carboniferous forests of 300 million years ago. Both specimens were presented to the Museum by the National Coal Board.

These fossil leaves of an extinct seed-fern (*Pachypteris papillosa*), although about 150 million years old, are in a quite remarkable state of preservation. They come from Middle Jurassic rocks of North Yorkshire and the deposit is so concentrated and the leaves so robust that they can be picked out and mounted on cards (as shown here). They were collected in about 1912 by Hugh Hamshaw Thomas from the classic locality at Roseberry Topping near Great Ayton. Thomas apparently delighted his friends by sending similar specimens on cards at Christmas time. The species is one of the most interesting in this classic Jurassic flora because it is always associated with marine microfossils, suggesting that it may have been salt-tolerant like a mangrove. It probably colonized the seaward margin of the freshwater sediments deposited in North Yorkshire at that time.

All these shells belong to species of the West Indian land snail genus *Cerion*, which has long posed problems for the taxonomist because of the extreme variation in shell shape. As a result, more than 600 species names have been proposed within this single genus. The snails in life form populations of similar individuals which may be separated by only a few hundred metres from other populations representing an apparently quite distinct second species. Recently, however, it has been shown that biochemical differences between such populations are by no means abrupt, suggesting that much of the shell variation is merely a response to local conditions. As a result, the number of 'species' in the genus *Cerion* can probably be reduced a hundred-fold. This puts a very different interpretation on evolution within the genus. Although more than one species of *Cerion* is shown here, a similar degree of variation can occur also within a single species.

Neotrigonia is the sole survivor of the once large and diverse order Trigonoida, a group of marine bivalve molluscs that traces its ancestry back over 400 million years to the Silurian period. The few species of *Neotrigonia* are found only around Australia. They have a large muscular foot with which they can burrow rapidly and they live in much the same way as the European cockle (*Cerastoderma edule*), but down to depths of more than 100 metres. One of their primitive features is the lack of siphons, water being drawn into the mantle cavity (for feeding and breathing) by means of temporary apertures formed by the mantle lobes.

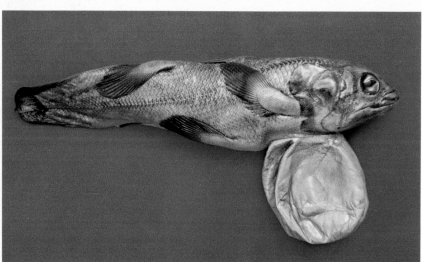

Ever since the discovery of a living coelacanth (*Latimeria chalumnae*) off South Africa in 1938 there has been enormous interest in the anatomy, physiology, and life style of these fishes, which until then were known only from fossils. Many more specimens have since been caught in the Indian Ocean, although only from one small area north of Madagascar. Quite recently dissection of a 1.6-metre female revealed five almost full-term embryos, of which one is now in this Museum. Unlike most fishes, therefore, the living coelacanth does not lay eggs but gives birth to fully formed young. Fossil coelacanths have a history going back some 300 million years but they disappear from the fossil record about 75 million years ago.

Over a long period of time the herring gulls have spread in a broad belt around the Arctic Ocean. As in many widely distributed animals and plants, differences have arisen between the various populations, but such populations interbreed where they meet and thus act like members of a single species. Before the end of the Ice Ages this 'ring' was not closed, however. Thereafter the pale-backed North American gulls crossed the Atlantic and overlapped the darker-backed form of northern Europe. Since these two do not interbreed, they are considered to be separate species, the herring gull (*Larus argentatus*) and the lesser black-backed gull (*L. fuscus*), even though the two merge imperceptibly throughout the rest of the ring.

One of the cornerstones in Darwin's argument in favour of evolution through natural selection was that if variation could be made to occur in domestic animals and plants, then the variation seen in nature must have the same cause, in other words, a form of selection. Although this does not prove the argument, the analogy is a sound one. Darwin carried on a large correspondence with breeders, including pigeon fanciers, and the Museum has study skins and skeletons from his own collection of pigeons (which do not show the differences so strikingly as the mounted specimens here). In 1868 Darwin published his conclusions in his *Variation of animals and plants under domestication*.

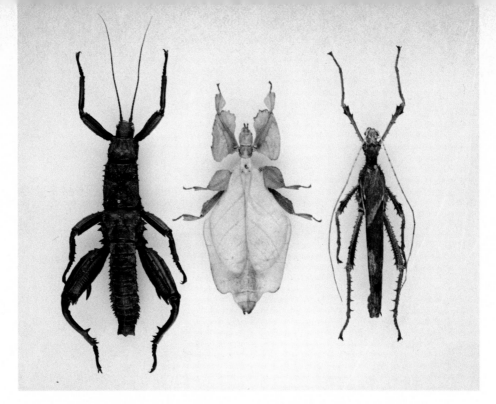

Although it is sometimes fairly obvious how small variations in individuals within a population can be advantageous and thus be passed on to future generations, eventually to become the norm for all members, the steps by which some of the more bizarre adaptations have been achieved can only be guessed at. This is particularly the case in mimicry. Three extreme examples of camouflage are shown here. The oriental leaf-insect (*Phyllium pulchrifolium*) is a striking case of leaf-resemblance (*centre*), while the two stick-insects are equipped with sharp spines that are used in defence should the camouflage be detected (*left, Eurycantha calcarata* from New Guinea; *right, Heteropteryx dilatata* from Malaysia).

One of the classic examples of the way that natural selection, the mechanism of evolution, can actually be seen at work is in the peppered moth (*Biston betularia*). Until the mid-nineteenth century and the advent of industrialization in the Midlands of Britain, the moth existed in its typical grey form which makes it almost invisible on lichen-covered trees. As tree trunks became increasingly soot-blackened, a dark melanic form of the moth appeared, the result of a single gene mutation. In 50 years this came to dominate the population, simply because the grey form was no longer camouflaged against predation by birds. With modern restrictions on pollution, the melanic form seems to have decreased, although other factors may be at work. The specimens illustrated are from the collection of H. B. D. Kettlewell, who did the initial research on this fascinating subject.

Certain butterflies, whose caterpillars feed on plants poisonous to vertebrates, retain this poison and use it for their own defence, advertising the fact by their bright 'warning' colouration. Through bitter experience, would-be predator species learn the meaning of these bold colour patterns and avoid such butterflies on sight. Other butterflies, which are palatable and often belong to quite different families, have evolved remarkable resemblances to particular poisonous species and are thought to gain protection as 'sheep in wolves' clothing'. For example, the palatable satyrid butterflies (*top row*), whose caterpillars feed on grasses, are normally cryptic, brown-coloured insects. On Madagascar an extremely rare satyrid (*middle row, left*) mimics a poisonous pierid butterfly (*below it*), whose caterpillars feed on mistletoe. Even more remarkable is a rare satyrid from New Guinea in which the male (*middle row, right*) resembles a poisonous ithomiid butterfly (*below it*), whereas the female (*middle row, centre*) is like an amathusiid butterfly (*below it*).

Many aspects of evolution are difficult, perhaps impossible, to prove by way of experiment, but some confidence can be felt if they lead to predictions that can be confirmed. Charles Darwin predicted, as a result of his theories, that a moth must have evolved in Madagascar that had a tongue some 20 centimetres long, since such a tongue would be needed to reach the end of the nectar-producing spurs of the Madagascan orchid *Angraecum sesquipedale* (*see inset*). Prediction became fact when the hawkmoth *Xanthopan morgani praedicta* was discovered, with a tongue long enough to reach to the base of the slender spurs.

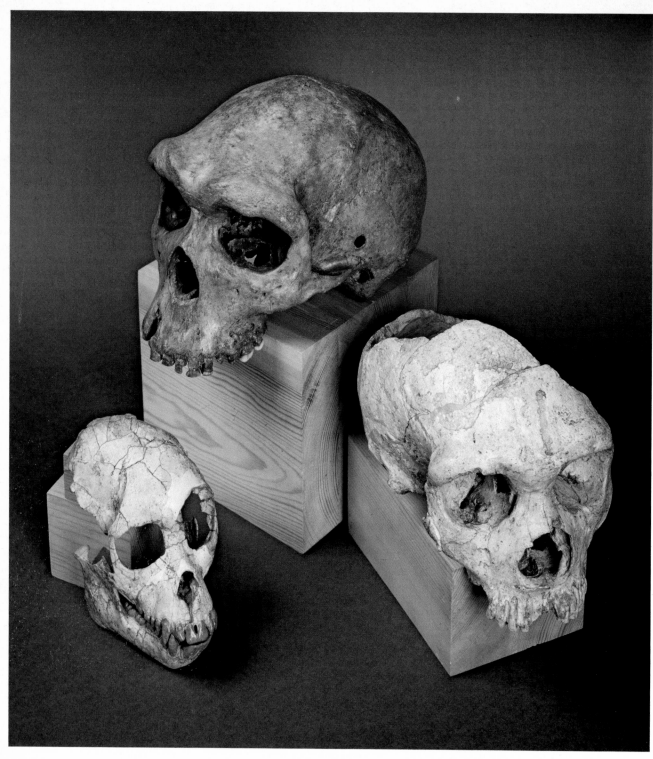

Perhaps the three most celebrated skulls in the Museum collection are *Proconsul* (*left*), Rhodesian Man (*centre*), and the first discovered skull of Neandertal type (*right*). *Proconsul africanus*, discovered by Mary Leakey at Rusinga Island in Kenya in 1948, was the size of a baboon and shares characters with both apes and man, although it lived some 18 million years ago and thus well before these two stems diverged. The beautifully preserved skull of Rhodesian Man, perhaps 120 000 years old, was found at the Broken Hill Mine at Kabwe (Zambia) in 1921 and although formerly considered a distinct species or even genus, is now recognized as our own species, *Homo sapiens*, in spite of the tremendous brow ridges. The Gibraltar skull, from a Neandertal woman of about 70 000 years ago, was discovered at Forbes' Quarry in Gibraltar in 1848, unfortunately without any other associated bones or artefacts.

Natural history and human history

The study of human history might seem to be the province of the archaeologist and historian, but in fact natural history can contribute a great deal toward understanding our past. Indeed, in one sense human history *is* natural history, because human societies are as much a part of the natural world as any other animal or plant populations. The physical aspects of human evolution are explored by physical anthropology, for which there has been a separate Sub-Department in the Museum since 1959; it also joins with palaeontology in the search for our pre-human origins. Human history took place in an environment with particular sets of climatic conditions and particular floras and faunas; to reconstruct these, specialists at the Museum are frequently called upon to identify animal, plant, or mineral remains. Similarly our cultural past cannot be properly reconstructed without identification of the material clues that have been left. Furthermore, natural history as a practical activity or as a mature science has been a social activity from earliest times, so that a knowledge of the history of natural history throws considerable light on the way that our societies have developed. Finally, the history of the way that mankind has used natural resources is essential to an understanding of the choices now open to us; once again, this is a task for naturalists.

In the twenty years that preceded the opening of the Museum at South Kensington in 1881, controversy over the antiquity of *Homo sapiens* and our relationships to the higher apes was at its height. The first Neandertal remains had been found as early as 1848, but almost all of the other important links in understanding human evolution were yet to be discovered. The evidence, however, was enough for Darwin's most loyal supporter Thomas Huxley to accuse Richard Owen of distorting anatomical truth to support his belief in the special creation of man. (Curiously enough, on the very spot where Huxley unveiled the statue of Darwin in the Main Hall in 1895, now stands the statue of Richard Owen.) There are today three veritable milestones in the Museum's collections: the celebrated skulls of Rhodesian man, of a Neandertal women, and of the early ape *Proconsul africanus*. More humble remains, however, are also of interest. For example, the worn-down molar teeth from certain Saxon burials tell us that the bread contained a great deal of grit from the stone used to grind the corn. Again, a study of bone lesions in pre-Columbian or Polynesian skeletons can help toward solving the old problem of whether syphilis was brought back to Europe by Colombus' expeditions or later was spread to innocent Tahitians by Captain Cook's crews.

A major preoccupation in early societies was with the seasons and with climatic change (or the climatic differences that were experienced in large-scale migrations). In trying to deduce past climates or weather patterns, the help of natural historians can be critical. Usually this means identifying the plants and animals associated with archaeological sites, coupled with a good knowledge of the present habits and environmental requirements of the species involved. For this type of study pollen grains are ideal. Small, numerous, and widespread, their inert and resistant shells enable pollen to be recovered in the laboratory from fossil deposits and to be separated from the peat or silt in which the grains are embedded. From them can be reconstructed the type of vegetation and its changes through successive layers as the climate altered or forests were cleared to make way for agriculture. In the same way, bones or hard parts of animals can give a great deal of information on early environments, as for example in distinguishing between warm and cold periods during the Ice Ages. The woolly mammoth discovered at Aveley in Essex in 1964 overlay the skeleton of a straight-tusked elephant, the first an animal of the cold phases of the Ice Age, the second of the warm interglacial periods. However, plant remains associated with the skeletons showed a change from milder woodland conditions to a more sparse and open vegetation, thus confirming a change of climate. The Aveley skeletons are now on public display.

Tool-making is a hallmark of our species, made possible by the co-ordination of the hand and eye as a learned technique, and it received its greatest impetus when language became the means of transmitting the skill. Identification of the materials used for implements can often throw interesting light on such questions as trade routes if the stone or other material came from a distant source; or on the type of animals present, leading to problems of how they were caught, which in turn may explain the use to which some puzzling weapon was put. All manner of implements have been brought to the Museum to be identified and the natural history aspects of their material have greatly helped archaeological interpretation. In addition to implements, all the other artefacts of societies at different, often less-advanced levels of culture, such as clothing, decoration, religious symbols, or art works, have also been fashioned from some part of the natural world and require to be identified and their materials understood in the context of human activities.

The Museum is often asked to identify animals or plants in works of art, and from the subjects in paintings and sculpture a great deal can be deduced of past knowledge and interactions with the natural world. The exact identification may also be of great interest to the art historian in tracing the artist's sources or even in dating if the subject is an exotic species whose date of introduction is known. Life studies of animals and plants can often throw light on the artist's movements; in one case the 'hippopotamus' in a painting was shown to be a capybara, a large rodent found in Brazil where the artist in fact worked, thus saving him a quite mythical journey to Africa. In turn, the natural historian can learn about agricultural or horticultural practices and the stages by which modern cultivated plants have reached their present form (or about steps in the breeding of domestic animals). Natural history drawings and paintings sometimes combine both art and science; for the art historian they may provide the primary source for tracing the advent of exoticism in a particular period, whereas for the naturalist they will perhaps enable him to identify species otherwise known from curt Latin descriptions. Again, the opinion of a specialist on the accuracy of the representation can be interesting. Certain Chinese drawings of insects of the nineteenth century might be marvelled at for the sheer observational powers of the artist, but the entomologist will point out that some of the insects are pure invention, which thus leads to revised ideas on the intention of the pictures and their place in that particular society. In this way, the collaboration of the natural historian with specialists in other fields is of mutual benefit in trying to arrive at a more detailed understanding of human history.

New scientific ideas not only arise and develop within the context of quite definite social attitudes, but in turn they have their own particular ways of redefining those attitudes. Once again, this is a field to which the natural historian can contribute through his understanding of at least one part of this two-way process. Darwin's *Origin of species*, in which he did no more than hint that 'light would be thrown on the origin of man and his history', led to an immediate furore. Sixteen years later, however, when Darwin finally published his *Descent of man*, many naturalists had accepted his thesis that human descent was from other species, but outside of scientific circles the arguments that raged affected wide areas of Victorian thinking. In order to understand the profound revolution in ideas, it is important that natural historians should be able to provide accurate scientific details of the case, often using their special knowledge of the subject to decipher or draw attention to scraps of information which the historian may pass over. Darwin has probably attracted more attention from writers than any other biologist, but there are thousands of other naturalists whose contribution to the evolution of culture requires appraisal by their fellow workers.

Yet another important aspect is to try to understand how the aims of natural history have arisen from particular social needs and have received sanction from society. The naturalist of means, such as Sir Hans Sloane, Sir Joseph Banks, or Lord Rothschild (all of whom are remembered as benefactors of the Museum), is rare in any generation and certainly nowadays natural history is an institutional activity that requires considerable financial support. An appreciation of the way that naturalists have come upon their problems and the kind of support their society has given them can throw light on the way that this happens (or ought to happen) nowadays. The extent to which naturalists are directed or allowed to roam free (or are merely nudged from time to time) is a problem with historical roots buried deep in the way that society sees its relationship with the environment.

The most crucial issue in the modern world (in fact a multiplicity of issues) concerns the management of our natural resources, whether this is seen in terms of the squandering of irreplaceable minerals or the impossibility of growing sufficient food by conventional means to keep pace with population growth. This is as much an issue for natural history as it is for other spheres. It did not arise overnight but has a history of great complexity, a part of which is best understood by natural historians.

Legend has it that over a thousand years ago the Vikings were able to navigate their ships in northern oceans with the help of a 'sunstone' that indicated the position of the sun when it was hidden by clouds. None is known to have survived, but modern knowledge of minerals suggests that cordierite was probably the one used. This mineral changes colour when it is turned around while being observed in plane-polarized light (light waves vibrating in one direction). The colour change can be seen when the mineral is viewed against a small area of blue sky, the light scattered from the blue being polarized with the direction of vibration at 90° to the sun. Presumably, this was the technique used by the Vikings when only a small patch of blue sky was visible.

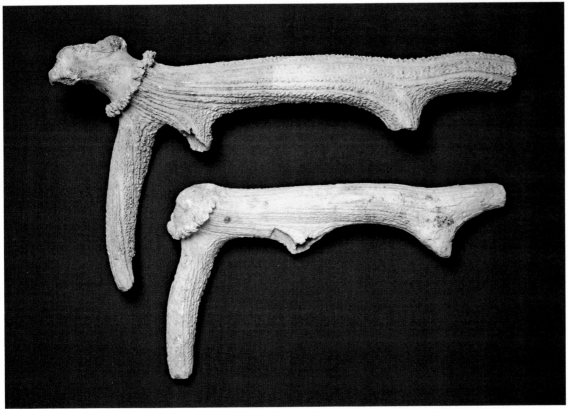

The ancestor of the modern pickaxe was the antler, cut to shape and leaving the brow tine and the main beam; similarly, the ancestral shovel was an animal scapula or shoulder bone. The antler picks shown here were from Grimes Graves, a Neolithic site in Norfolk, where the best black flint could be excavated from the chalk. Perhaps disappointed at not locating a flint seam in one shaft, the men heaped up chalk into a rought altar, placed a fat little goddess on top, and left their picks in a pile, to be discovered many centuries later.

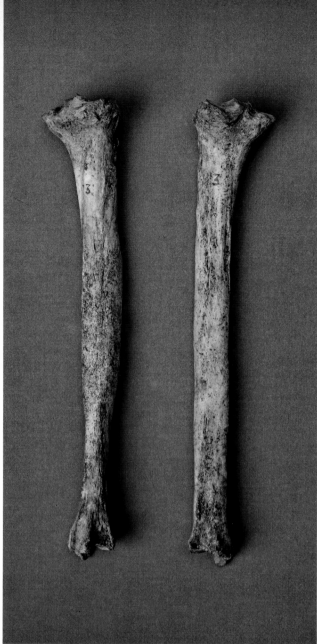

This pre-Columbian trepanned skull from the Cuzco region of Peru, rather irreverently known here as 'Holy Joe', has no less than seven trepanned holes in it, all probably made with stone implements (up to eleven holes have been recorded, however). Why a patient endured so many operations without modern anaesthetics is a puzzle, but it suggests some recurrent affliction like migraine or epilepsy. Usually only a single hole is found. Peru has produced most cases of this operation, but it has been widely and independently practised in a variety of cultures and at various periods, with examples from New Britain in Melanesia, North Africa, France, and Britain. Freshly flaked stone tools are relatively clean; the operation fell into disrepute with the introduction of metal. Many clues to the early history of diseases can be deduced from skeletal material.

Ever since the dramatic appearance of syphilis in Naples in 1495, Europe has blamed America for the introduction of this disease, but more recently the examination of 4500 pre-Columbian American Indian skulls showed that less than 0.2 per cent had any sign of syphilis. The picture is complicated, however, because the bacteria (treponemes) responsible for yaws, a related disease, are indistinguishable from those of syphilis. Is one an urban and the other a rural manifestation of the same disease? Had immunity been achieved on each side of the Atlantic, to be exploded when the two strains met in Naples? Bones from archaeological sites can help toward resolving such questions. Syphilis is the only disease with a poetic name, being derived from Syphilus a handsome shepherd whose misfortunes (and the supposed cure) were recounted in a Latin poem published in Verona in 1530

The Furze Platt hand-axe was found by Mr G. Carter in March 1919 during his work in the Cannoncourt Farm Pit in the village of Furze Platt near Maidenhead. It is the largest hand-axe yet found in Britain, perhaps the largest in Europe. It is 30.6 centimetres long, 15.4 centimetres wide, and 5.4 centimetres thick; it weights 2.8 kilograms and would have required considerable strength to use. The edges are thin, but scale-like flaking was applied to make it still sharper. Some authorities, however, think that it was not used but merely treasured for its size and beauty. It comes from the Middle Acheulian culture and may be 250 000 years old. Perhaps some Acheulian stoneworker, impressed by the size of this flint nodule, decided to express his knapping skill without thought for the use of such a huge object.

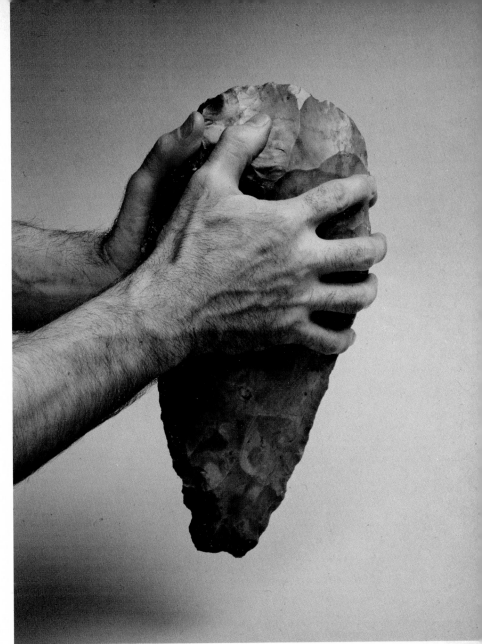

Ever since the development of tools with metal cutting edges, from knives to scythes, it has been necessary to sharpen them on smooth pieces of rock known as hones. Such hones are fairly common in archaeological sites from the Iron Age onwards. The large hone shown here (*top*) dates from the twelfth century and was excavated at Winchester. Perhaps surprisingly, it exactly matches rock specimens of schist from the known honestone quarries at Eidsborg in Norway (*bottom left*). The smaller hone (*bottom right*), also from Winchester, dates from the eleventh century and is of sandstone that undoubtedly came from the Bristol area. Specimens like these can thus give much useful information on trade patterns in former times.

Cowry money has been used at one time or another by many cultures, from West Africa to the Central Pacific, in some places as a short-lived phenomenon but in others continuing up to the beginning of the present century. Two species of cowry were used, *Cypraea moneta* and *C. annulus*, both of them common and widely distributed throughout the Indo-Pacific region; for the West African cultures they were therefore imported, which perhaps increased their value. Cowries are handy, lasting, easy to count, and difficult to counterfeit, thus fulfilling most of the properties of money. In China they were in use from about 1000 BC and the written ideogram for 'wealth', 'precious', and 'to buy' incorporates the symbol for cowry. Their most extensive use was in India and the main source for them was the Maldive Islands, where they were farmed by putting coconut branches into the water, thus attracting them to a suitable habitat.

From earliest times, shells have been used for decoration, either on the person or on objects. This shell mosaic or shadow box is Victorian, an age when fussy decoration was sometimes beautiful but often, to our eyes, absurd. Here, various gastropod and bivalve shells, as well as that of the cephalopod *Spirula*, have been worked into a pleasing pattern, the box presumably covered originally with glass and hung as a wall ornament. Such Victorian mosaics are similar to the 'sailors' Valentines' produced in the West Indies at the same period, of various shapes but often hinged pairs of octagonal shadow boxes that would open like a book; the themes would include Cupid's darts, hearts, and some sentimental phrase. The identity of shells can sometimes be of great use for ethnologists tracing the provenance of decorative objects from primitive cultures.

The skull of the crucifix fish (*Arius*) still has religious associations for many people in the Caribbean area, although the elaborately painted skulls are largely produced as tourist curios nowadays. Seen from above, the surface of the skull of these catfishes resembles a cloaked priest with outstretched arms. From below and when suitably decorated, the bones forming the cranial superstructure bear a remarkable likeness to a crucifix, the figure of Christ being delineated by paint. Each part of the skull represents an instrument of the Passion, as for example the otoliths or earstones, which rattle when the skull is shaken and are likened to the dice cast for Christ's garments. Although not of direct interest to taxonomy, such objects are a useful reminder of the many ways in which man uses the natural world.

Until fairly recently, crocodiles were regularly found throughout the length of the Nile, extending right into the delta. They played an important part in the life of ancient Egypt, in some places being hunted (as shown in the wall paintings decorating the tomb of Pta Hotep at Sakkarah), but in others being venerated. They were often mummified and placed in graves and many such mummies are known from around the temple of the crocodile god Sobk at Kom Ombo. The origin of the one shown here is unfortunately not known, but many such curiosities were brought back to England from at least the eighteenth century. At some time it lost the bandaging that evidently enclosed it and most of its legs are also missing.

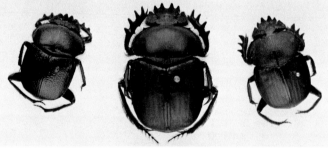

Perhaps the most influential of all beetles on the culture and mythology of a civilization is the sacred beetle (*Scarabaeus sacer*), the largest of the three shown here, whose symbolic importance in ancient Egypt dates back some 5000 years. It was seen to roll balls of dung and bury them, while every year beetles would emerge from the ground. Similarly, the sun god Khepri was seen as a great scarab rolling the sun across the heavens and every day the sun would be reborn. The scarab thus became a symbol of rebirth in the same form after death and their sculptured likenesses were put into tombs from the earliest dynasties to ensure the soul's immortality. Later, scarabs were worn as amulets of good luck. Of the 19 000 described species in the family Scarabaeidae, not all are 'dung beetles', but those that feed on dung are of great economic interest to countries like Australia where the dung of introduced domestic animals is largely left unburied by the native insects, thus lowering pasture yields. The Museum's collections help research workers to find suitable species to combat this problem.

Over a hundred axes made from jadeite and similar minerals have been found in Britain, while very many others have been discovered in France, Germany, Switzerland, Austria, and Italy. All belong to the Neolithic period, two having been dated as 3200 and 2860 BC, and larger axes in particular are finely polished and in perfect condition, indicating a very high degree of craftsmanship. Unfortunately, the jadeite source for these superb implements is not yet known, but work in the Museum on the mineralogy and chemistry of the materials may supply a clue and thus throw further light on the apparently extensive trade routes used by the Neolithic people. The axes shown here are from the collections of the British Museum (Bloomsbury).

The other scarabs shown here are *Scarabaeus sanctus* of India (*left*) and *S. festivus* of East Africa (*right*).

The Museum can often give considerable help to archaeologists in identifying plant and animal remains from sites of human occupation. The toothed pharyngeals or throat-bones of the corkwing wrasse (*Crenilabrus melops*) found in a Neolithic chambered tomb at Quanterness (Orkney) suggested that this species was much more abundant then than it is now. Since this fish has a southern distribution, its comparative abundance in Orkney seems to imply that the climate in Neolithic times was much warmer. Measurements on the bones and comparison with modern material gave an estimate of the size of the fishes caught, and in turn this led to consideration of the fishing methods available to such people and thus the level of their technology.

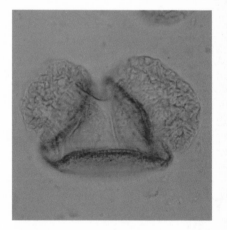

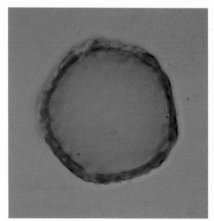

In fossil deposits or at archaeological sites pollen grains are often better preserved than other parts of plants, so that much of what we know of the countryside inhabited by our forebears and the effects that they had on it comes from the accurate identification of pollen. In this way we can chart the beginnings of agriculture by looking for the earliest occurrence of the pollen of crop plants or an increase in the weeds associated with cultivation, such as the sow thistle or plantain. As agriculture increased, forests were cut down and one finds fewer pollen grains from trees like pine or elm. The Museum keeps a large collection of microscope slides and electron microscope photographs of modern pollen grains for comparative work. Shown here are scotch pine (*Pinus sylvestris, top*) and wych elm (*Ulmus glabra, bottom*).

This shell, now rather battered, came from the tortoise that belonged to the celebrated amateur naturalist Gilbert White (1720–93), whose classic book *The natural history and antiquities of Selborne* (1789) is a fine example of patient and careful observation of nature. The tortoise, named Timothy, belonged to a woman who lived at Rigmer near Lewes (Sussex), who had kept it for some thirty years until March 1780 when White acquired it and brought it to his home at Selbourne (where it lived another fourteen years). In his book, which is largely a series of letters written to his fellow naturalists Thomas Pennant and Daines Barrington, White makes several references to Timothy, describing his preparations for hibernation, food preferences, and ability to predict rain. It was probably a spur-thighed tortoise of the species *Testudo graeca*, found from Greece, through Turkey, to northern Iran and northern Africa.

Mick the Miller

For about 10 000 years the well-being and at times the sheer survival of much of mankind has depended on the skill with which domesticated plants and animals could be manipulated by breeding to produce a higher yield or to adapt to different conditions. Such skills, which produced the shire horse and the merino sheep, have also been turned to leisure activities and the greyhound Mick the Miller must represent the acme of such techniques. Bred in Ireland by Father Brophy, he had contested 48 races by 1931 and had succeeded in winning 36 of them, including four classics, a record unbeaten until the appearance of Westpark Mustard in the 1970s. Mick the Miller's body was presented to the Museum in 1938 and is a fitting symbol of the degree to which natural form can be moulded to human interests.

In the wild state the goldfish (*Carassius auratus*) is dull brown, but at least 2000 years ago the Chinese discovered that unusual mutant forms could be bred. The first book on goldfish culture was by Chang Ch'ien-te (1577–1643) and from about this time specimens were brought back to Europe. Shown here is a page of drawings by a Chinese artist, made in Canton in the early part of the nineteenth century under the supervision of John Reeves (1774–1856), Inspector of Teas for the East India Company. As an amateur naturalist, Reeves commissioned several thousand such drawings of animals and plants in the Canton area, largely to indicate what species might be useful to Europe: this collection of drawings was eventually donated to the Museum. Reeves' contribution to English horticulture was immense, for through him came the first shipments of Chinese azaleas, camellias, peonies, and chrysanthemums.

From South America to the Pacific islands, colourful bird feathers have traditionally been used for decorative purposes, such as ceremonial clothes or to enhance weapons or ritual objects. In Hawaii at the time of its discovery by Captain Cook in the late eighteenth century, a number of indigenous birds were exploited in this way, including the I'iwi (*Vestiaria coccinea*), whose red feathers were much prized. The specimen shown here, preserved in alcohol but with the red still faintly apparent, was collected on Cook's third voyage and by a complicated route (involving Sir Joseph Banks and later the Royal College of Surgeons) eventually came to this Museum. The I'iwi is still found in Hawaii, but other birds formerly used for feather cloaks and helmets are now extinct, so that museum specimens are of considerable interest to ethnologists.

Sitodium . altile.

Otaheite

Sydney Parkinson pinxt 1769.

Between 1780 and 1787 over 15 000 slaves died from disease and starvation in the West Indies. To ameliorate this situation, Sir Joseph Banks proposed to George III the introduction of the breadfruit tree (*Artocarpus incisus*) from Tahiti. As a result, Captain Bligh was dispatched in the *Bounty* in 1787 to bring rooted suckers to the West Indies, but as a result of the famous mutiny it was not until 1791 that a second expedition succeeded and the breadfruit has since been propagated in the West Indies. Banks' suggestion was prompted by his experiences on Captain Cook's first voyage (1768–71) and it was on this voyage that the official natural history artist Sydney Parkinson (*c*. 1745–71) made the drawing shown here. The breadfruit was first mentioned by sixteenth-century Pacific voyagers, who found it widespread and of great importance to the Polynesians.

Using the natural world

In the early chapters of human pre-history, the activities of hunting, fishing, and gathering were a daily or seasonal search for what was traditionally useful. The range of animals and plants eaten or used was perhaps wide, but the utilization of new items most likely depended mainly on migration to new areas. With cultivation and domestication came a concentration on just a few animal and plant species and although much must still have been hunted or gathered, the impetus to exploit further elements of the environment will have stemmed chiefly from advancing technologies and cultural needs. Gradually the range of exploited species increased, but with human populations relatively small and their needs fairly simple, it appeared that 'nature' on the whole was bountiful, an attitude that persisted right into our own century. Some reserves might show signs of over-exploitation or even exhaustion, but in time they would recover or new sources would be found. Our outlook today is very different and those engaged in research at the Museum are fully conscious that many of the plants and animals that they deal with are threatened, that some are now extinct, that mineral reserves are irreplaceable, and that one of the outcomes of using the natural world is to change it, perhaps irrevocably.

Thus natural history, or the study of all aspects and relationships in the animal, plant, and mineral kingdoms, can never be an end in itself. At times it may seem to pursue goals little connected with our immediate needs, as for example when a scientist spends his working life to produce a monograph on some obscure group of insects or a little-known family of plants; another may devote considerable time to theories about classification, to tangles of nomenclature, to bibliographic problems, or to the sheer joy of just 'finding out'. However, natural history, no less than the world it explores, is a complex weaving, each part dependent on the others and no threads wasted. Each plant, animal, and mineral, however obscure, belongs to a web of ecological circumstances, and no such web exists with which mankind is not at some point involved. Similarly, all facets of natural history are ultimately joined by the overriding aim—to understand the natural world in order to manage it and to use it.

From the rise of museums in Renaissance times, the practical role of natural history usually found emphasis in museum catalogues (and presumably on the labels to the specimens themselves). The use made of an animal or plant was spelled out and there were always those ready to find some application for it in Europe, sometimes merely to create a new fashion, sometimes as a means of making wealth, and sometimes with a genuine desire to alleviate suffering or to enhance the lives of other people. By the beginning of the seventeenth century a stream of plants and seeds was flowing into Europe, as well as certain exotic animals. It is perhaps no coincidence that John Tradescant, the founder of what was to become the first public museum in England, was by profession a gardener (eventually to Charles I) and that in his own garden in South Lambeth he was the first to grow numerous plants from North America and elsewhere.

The beginning of the seventeenth century also saw the rise of the great East (and West) India Companies. Although primarily incorporated to exploit the wealth of existing resources, such as spices and tea in the East and sugar in the West, they did not neglect the search for new products. The men employed by the Companies were encouraged to explore the natural resources of the country in which they worked and often this went far beyond just the search for useful species. For example, Major-General Hardwicke (1756–1835), one of the Company's officers before India was handed over to the Crown following the Mutiny of 1857, made a huge collection of drawings of Indian animals. What is equally significant is that this collection (as well as specimens) was bequeathed to the British Museum, where J. E. Gray had already named and published descriptions of 200 of them in 1830–5. The earliest extant herbarium of Chinese plants (also now in the Museum) is that by a surgeon to the East India Company, James Cunninghame (*d*. 1709), who sent home some 600 species and about 800 drawings and who provided the first description of cultivated tea. Later, John Reeves, the Company's Inspector of Teas at Canton, set about commissioning Chinese artists to illustrate all the plants and animals on which he could lay his hands and some 2000 of these drawings are now in the Museum. However, the most celebrated figure in this search for useful overseas products was undoubtedly Sir Joseph Banks, whose participation as naturalist on the first of Captain Cook's three voyages had stimulated his interest in exotic plants. It was through him that Captain Bligh was dispatched on the ill-fated voyage of the *Bounty*. While the events of the voyage are well known, it is usually forgotten what the purpose of the voyage was—to accomplish Banks' suggestion of bringing from Tahiti rooted suckers of the breadfruit tree in order to provide an easily grown and harvested food for slaves on the West Indian sugar plantations. In 1803 Banks sent out his first botanical collector to China, the Kew gardener William Kerr (*d*. 1814); few of Kerr's plants survived the long journey home, but one of them nursed back to vigorous life at Kew was the familiar *Kerria*, named in his honour.

Other countries, with their own India Companies, were no less enthusiastic in the hunt for new products and as in England there was a close relationship between the collectors and the museums, promoted by wealthy individuals or by governments. During a period when the British Museum was in the doldrums, it was to Joseph Banks' own house in Soho Square in London (almost a prototype of the modern research institute) that naturalists from many countries came to examine material, to compare their own, and to discuss with other naturalists. Museums acted as a kind of focal point and much that was discussed was of a practical nature, as for example the uses of the cochineal insect, the composition of Chinese ink, the properties of new flowers and vegetables, the development of the sisal and other industries, or the possibility of acclimatizing exotic fruits. In conjunction with botanic gardens and zoos, museums came to be regarded as sources for information about new as well as existing resources, and increasingly it was to the British Museum that material from British expeditions was sent. As a result, the huge displays of animals, minerals, and plants in the early days of the British Museum and later at South Kensington were not only to show just the diversity of the natural world, but also—shades of bountiful nature—to give an idea of its richness in terms of human uses.

The properties of minerals can to a large extent be explored within the Museum, but the living properties of plants and animals are better studied at special institutes (although anatomical studies and field work by Museum staff can contribute a great deal). Essentially, what a museum supplies is identifications, clues to relationships, and an indication of how to set about using the literature. These three categories of information represent the groundwork on which all further studies are based. Without them, and especially without correct identifications, a great deal of biological and mineralogical work would come to a halt or produce hopelessly misleading conclusions. In the determination of the relative ages of rocks by means of fossils, in analysing a tropical fishery, or in judging the effects of forest clearance, nothing can be certain until the elements have been properly identified. Thus a field worker may feel it safe to exploit a particular animal even during its spring period of reproduction since it also breeds again in the autumn; the taxonomist, however, may point out that there are in fact two quite separate species and the first could be seriously threatened by such exploitation.

One of the most important contributions made by the staff at the Museum is their help in compiling accounts of the fauna and flora of various regions. A most practical example is the volumes of identification sheets for fishes of economic or subsistence value in the major fishing areas of the world, organized by the Food and Agriculture Organization of the United Nations and intended as a guide for fishery workers. Such large-scale works can now be produced only through the collaboration of numerous specialists, often spread over several museums and institutions. In this way, the years that a specialist may have spent in studying some particular group can suddenly be put to very practical use, whereas his scientific publications have hitherto been mainly of value to his fellow workers.

The range of inquiries and the studies that they have provoked within the Museum are as varied as the uses to which the natural world has been put. They include the culture of aquatic animals, the production of silk, horticulture and the use of herbs, fisheries, oil exploration, livestock breeding, drugs, the decorative arts, whaling, and the fight to preserve threatened species. It is noticeable that increasingly the days of simply using the natural world like some giant cornucopia are going and that problems brought to the Museum usually contain an element of concern for future prospects. Almost unnoticed, the role of the Museum is being transformed, from that of an active agent in the exploitation of natural resources, to that of the cautioning adviser on their management.

The culture of prawns, and especially the large species such as *Penaeus indicus* and *P. monodon*, is an important aspect of fisheries in many tropical countries. Instead of waiting for the juveniles to invade ponds at high tides, culture farms are now breeding their own stock, which requires considerable knowledge of the biology of the specie concerned. Successful methods for laboratory rearing of prawns and shrimps have been developed in the Museum and various aspects of their life histories have been studied, in particular larval behaviour. In turn, this has added to knowledge of the relationships of species and their evolutionary history. *Inset*: a stage-5 larva of the ditch shrimp *Palaemonetes varians* prior to metamorphosis.

The various species of krill (shrimp-like euphausids) constitute the major part of the diet of whales in Antarctic waters. These euphausids are so abundant that a moderate-sized blue whale may consume two or three tonnes of them at each meal. The economic importance of krill, both in relation to the whale fisheries and as a possible fishery in itself (to be reduced to a protein flour), has led to an intensive study of their biology. The Museum's comprehensive reference collection and library resources are frequently used by biologists who need to check the identification of their stud material. Shown here is *Euphausia superba*.

The earliest known text on medicinal plants is the *De materia medica* by the Greek physician Dioscorides, written in AD 77–8. It was a manuscript that was very widely copied and used, being the standard medical work up to and throughout the Middle Ages. The most famous manuscript copy is the superbly illustrated one in the National Library in Vienna, which was made in about AD 492 for Juliana Anicia, daughter of Flavius Anicius Olybrius, Emperor of the West. Another fine manuscript version, made in the 1460s, or a thousand years later, was once owned by Sir Joseph Banks and is now in this Museum. Compared with the wonderfully delicate plant drawings of Dürer or Leonardo da Vinci, made only a few decades later, those in the Banks volume are perhaps rather crude and naïve, but one of the best is the bramble shown here, entitled *Rubus sylvestris*.

The cochineal insect (*Dactylopius coccus*) is commercially the most important of nine sap-sucking insects (Hemiptera) that produce carminic acid, which after chemical processing forms the brilliant-red dye carmine. This is widely used to colour drinks, foodstuffs, and cosmetics. The cochineal insects are native to the Americas and live only on cacti, particularly the genus *Opuntia* (which includes the prickly pear). The insect of commerce was deliberately grown by both the Aztecs and the Incas and used for dyeing, and with the Spanish conquest production increased and spread to other parts of the world. The main areas nowadays are Peru and the Canary Islands. Identification of the nine species is extremely difficult and can only be done using microscopic characters. All nine species have also been used in the biological control of cacti since they cause damage to plant tissue.

Among the 2000 Chinese drawings of plants collected by John Reeves while Inspector of Teas in Canton, are many superb pictures of garden flowers which were mainly unknown in Europe. Among Reeves' drawings are about 60 varieties of chrysanthemum, some of which are shown here. Chrysanthemums brought to Europe from Japan in the seventeenth century did not survive, but a purple–crimson double-flowered cultivar from China was successfully grown in 1789. Reeves added very many further varieties, sent back from Canton in pots in the early decades of the nineteenth century; he was also responsible for introducing *Wisteria sinensis*. The origin of the florist's chrysanthemums (*Chrysanthemum morifolium*) is obscure, but part of its parentage is probably a small, yellow-flowered species (*C. indicum*) from China and Japan, whose cultivation in China dates back to 500 BC.

The use of silk as a natural textile fibre has been known to the Chinese for over 5000 years, although it was kept as a secret until the fourth century AD, when specimens of the mulberry silkworm moth (*Bombyx mori*) reached Europe. The moth probably originated in the Himalayas, but is no longer found in the wild state. The silk is unravelled simultaneously from several cocoons floating in near-boiling water, the fibres being reeled together to make a single thread of suitable thickness. From a single cocoon (shown here on the right) comes a thread of one kilometre or more in length. Since the mulberry silkworm moth cannot fly and its larvae will live in shallow trays without escaping, it is the only fully domesticated silkworm. Some members of the family Saturniidae also produce silk that can be reeled from the cocoon, but they must be cultivated in the wild; the most spectacular is the atlas moth (*Attacus atlas*), with a wingspan of about 25 centimetres, from whose immense cocoons is made Fagara silk.

Conservation measures have saved a number of rare breeds of livestock, but many of the old breeds are now extinct or have changed beyond all recognition. Knowledge of such breeds is important to understanding the history of breeding techniques, but it is altogether unusual to have such fine models of them as are shown here. They were made between 1790 and 1810 by George Garrard (1760–1826), a painter sponsored by the Board of Agriculture, who modelled at least 21 breeds and published two volumes of coloured engravings of cattle (but unfortunately not pigs and sheep) with precise measurements. The wild boar (*left*), by then extinct in Britain but introduced sometimes to parks, was ancestor to the old English boar (*centre*), which from the 1770s was crossed with the imported 'dish-faced' Asiatic pigs; the result of this cross (*right*) is an ancestor of our present-day breeds.

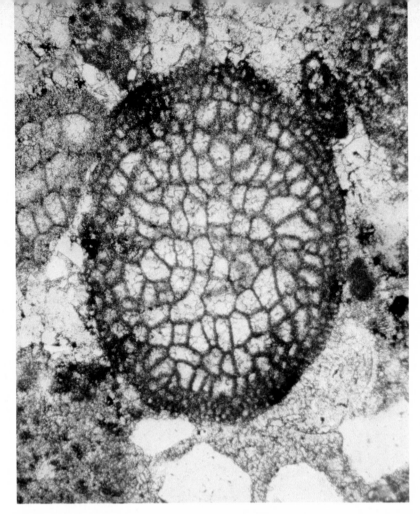

This delicate lace-like pattern of a fossil seaweed (*Subterraniphyllum*) is the result of deposition of lime within the tissues during life. When fossilized, such specimens can be cut as a thin transparent slice and the structure seen almost as clearly as in the living material. Species of this seaweed were first found in deep borings for petroleum in the Middle East, but are now known from many localities in the Mediterranean countries. Fossil remains such as this usually mark quite distinct levels in the rocks, so that accurate identification can provide a most useful tool in the exploration for oil and minerals. The one shown here came from rocks of Oligocene age (about 38–26 million years ago). The original specimens, from which the first description was made, are in the Museum collections.

Foraminifera are small single-celled animals (Protozoa) which secrete a shell, usually of calcium carbonate. Most are microscopic, but some like the ones shown here can grow comparatively large and are easily visible to the naked eye. These larger foraminifera live in shallow water and at various times in the geological past they evolved rapidly. They were also widely distributed in tropical regions, and in some places were so numerous that whole layers of rocks were built up from their shells. As a result they are extremely useful to geologists, especially in the oil industry, since from their known time range and distribution these larger foraminifera act as markers to date limestones over large areas of the globe. This slice of foraminiferal limestone contains members of two genera, *Discocyclina* and *Amphistegina*. It comes from Ecuador, and is part of the material studied in a joint project with the Overseas Division of the Institute of Geological Sciences in London. Almost all the research carried out by the Museum's Protozoa Section of the Palaeontology Department is connected with Overseas Geological Surveys.

Herrings, sardines, sprats, pilchards, and anchovies are the most important of all groups of exploited fishes, comprising a third of world catches and in particular areas making up half or more of the catch. The catches of anchovy off Peruvian coasts were at one time the highest for any single fish species in the world, reaching 12 million tonnes in 1970. In temperate waters, the number of species is few, but in tropical seas there may be 40 or 50 species available to the fisheries. Their great similarity to each other makes identification extremely difficult for the field biologist, yet no useful biological data or warnings of over-fishing can be learned if the species are misidentified or muddled. Shown here are *Sardinella gibbosa* (*top*) and *S. albella*, both from the Indo-Pacific region. Museum studies have shown them to be distinct but on the basis of characters that require a microscope; field studies may now reveal more obvious differences.

The collection of large whale skeletons, seen here before transfer to more spacious quarters in outer London, contains specimens from strandings or from earlier whaling days. They have been used for numerous studies as well as inquiries, either for identifying species from photographs and descriptions, or for biological, taxonomic, or anatomical research. The Museum has kept records of all species of cetaceans (whales, dolphins, porpoises) stranded on the coasts of Great Britain and Northern Ireland since 1913. These have mostly been the property of the Crown since 1324 and are normally reported to the Museum by H.M. Coastguards and Receivers of Wreck. Serious moves to protect whales began at the turn of the century, and in 1921 legislation for conservation began in earnest.

Mother-of-pearl or nacre forms the innermost shell layer of many molluscs. It has from earliest times been used for a variety of ornamental purposes, but perhaps most extensively for the manufacture of buttons. The shell shown here is a black-lipped pearl oyster (*Pinctada margaritifera*). Only certain shells, such as the ormers, top-shells, and pearl oysters, were sufficiently abundant and produced a thick enough nacreous shell layer to be used commercially. Pearl-button manufacture reached its peak in mid-Victorian times, when up to 2000 tonnes of shells were imported into Britain annually. Discs known as 'blanks' were cut from the shells, drilled, and then polished—largely by hand; this perhaps explains the rapid decline of the industry with the advent of cheap plastic alternatives.

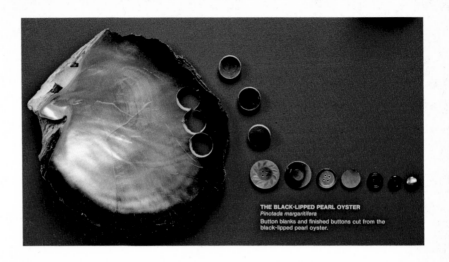

THE BLACK-LIPPED PEARL OYSTER
Pinctada margaritifera
Button blanks and finished buttons cut from the black-lipped pearl oyster.

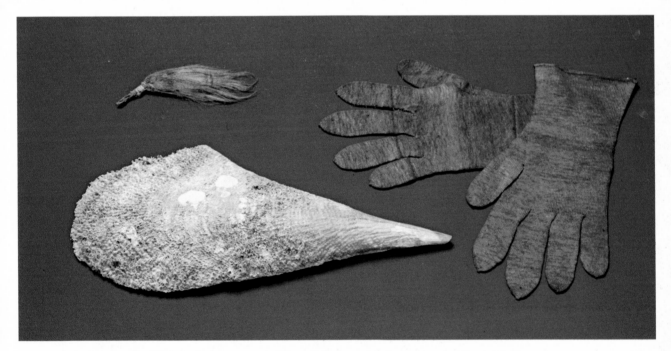

An unexpected source for fabric is the byssus or holdfast of the large Mediterranean noble pen shell (*Pinna nobilis*), a bivalve mollusc that lives in an upright position, apex downward, partially buried in sand or mud and rooted by its byssus. In clean sand the byssus is a glossy golden colour, but is dull and black in mud. To prepare the threads, the byssus is cut off, washed and half-dried, spread out, and allowed to dry fully. For fine work the threads are drawn through iron combs or cards and are then spun with a distaff and spindle, with two or three threads mixed with one of silk. From this are knitted gloves, like those shown here, stockings, or even whole garments. Those of a cinnamon or glossy gold colour are highly prized and it has been suggested that from such a garment came the story of the Golden Fleece. There was a thriving industry centred at Taranto in Italy in the eighteenth century, but only a few mainly tourist items are produced today.

This group of beautifully symmetrical crystals is composed of cassiterite (SnO_2), the most important ore of tin and one of the few tin minerals. It is from the famous collection of some 14 000 British mineral specimens assembled by Sir Arthur Russell (1878–1964), acquired by the Museum in 1964, and it came from the Polperro mine at St Agnes in Cornwall. Cornish tin-mining dates back to pre-historic times and the Cassiterides or Tin Islands of the ancient Greek and Roman geographers may well have been the Cornish peninsula. Cornish tin was probably the main source for Mediterranean countries during the Bronze age and was the world's principal source in the eighteenth and early nineteenth centuries. Tin ore is usually found massive or as granular or radiating intergrowths with other minerals; the prismatic crystals shown here are rare.

The quest for furs was probably one of the greatest stimulants for the exploration of Canada and Siberia, of which one product was the Hudson Bay Company, founded in 1670 and still Canada's biggest fur company. As a result, the fur-bearing mammals were among the first groups of animals to be collected for scientific study from these areas. Shown here are some naturally occurring varieties of the Eurasian red squirrel (*Sciurus vulgaris*). Some of these colour forms represent geographical races, while others coexist in the same area, the dominant form depending on the dominant species of tree. The Museum's large collection of fur-skins now aids customs officials to enforce legislation against trade in endangered species by enabling consignments of skins to be accurately identified.

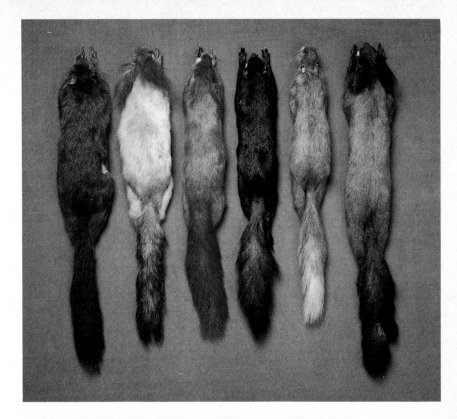

To combat excessive exploitation, some plants have developed chemical defences by synthesizing poisons to sabotage the digestive, nervous, circulatory, or growth systems of their animal enemies. These poisons far exceed in number and complexity the drugs manufactured by man. Fatal in large doses, some such plant products have proved valuable medical drugs when administered in minute doses. Some families, such as the periwinkles (Apocynaceae), are notoriously poisonous. The Madagascar periwinkle shown here (*Catharanthus roseus*, often called *Vinca rosea*) contains 60 or more slightly different poisons, three of which are now used in chemotherapy, and this has resulted in intensive biochemical research on the properties of related plants. Correct identifications are essential and the Museum has produced monographs on the genus *Vinca* or true periwinkles of Europe, North Africa, and western Asia, and on the tropical genus *Catharanthus* native to Madagascar and India, but nowadays grown elsewhere.

Ca.lrir.

Glandofa ⁊ Gnatrix. Conftanti
nus. Glandofa ferpens eft cuius
morfus eft putridus ⁊ de quo pe
ffimus odor eminus egreditur. Dicūt āt
quidam q̄ pedes fuper illam incedentis
excoriantur. ⁊ apoftema in crure paritur.
Illius quoq̄ qui ei medetur manus exco
riantur. Qui vo hunc ferpentē occiderit
odor eius putridus fit. ⁊ omē nifi folius
huius fetoris pdit. huius morfus fignifi
catio eft. q̄ in illo loco fubfequit̄ apoftea
cum rubore. vefica illa membra circūdat
aliq̄ fūmatim currit fanguis quafi aqua
dolorem quoq̄ patitur in ore ftōi. Gna
trix. Ifidorus. Gnatrix eft ferpens aquas
veneno inficiens. In quocunq̄ eū fonte
fuerit. cum veneno immifcet. De q̄ Luca
nus. Eft gnatrix violator aque.

Operationes.

Pli.li.rrr. Grillus prodeft p̄uulē
tis auribus cū terra fua effofus. magnam
huic aīali phibet nigidius. autoritatē ma
gi q̄ maiore. Auricule parotide profunt
grilli fiue illiti fiue adalligati. ⸿ H
Ideli.rrr. Strumis q̄ fuaderit illinire
grillū cū terra fua effofus: ⸿ Porro
⁊ calculos ⁊ alias difficultates vefice pro
deft grillū aq̄ calida diluti fume pterea
grillo₂ cremato₂ fauilia ex oleo perducit
ad cicatricem vlcera. Daly li. re
galis difpofitio. locufte longe.i. grilli. fi
collo fufpendant̄ q̄rtanam pacienti pro
deft. ⸿ Petrus comefto₂. Gurgulio
vermis eft qui nafcitur in corrupto͞e faba
rum ficut tinea. ex corruptio͞e olei. teredi
nes ex corruptio͞e ligno₂. Ifi. Gurgulio
dr. q̄ pene nibil eft aliud nifi guttur.

Ca.lrr.

Grillus ⁊ Gurgulio. Ifidorus
Grillus a fono vocis nomen ba
bet. hic retro ambulat. terram te
rebrat. ftridet noctibus. venatur cum for
mica circumligata capillo in cauernam
eius coniecta. efflato prius puluere ne fe
abfcondat. ⁊ ita formice pplexib⁹ trahit̄.

Ca.lrrj.

Ericius. Areftotiles:
Hericius eft animal fpi
nofum: cuius fenfus ap
paret multociens cōtra
vētos feptētrio͞alel ⁊ me
ridionales: foraīa nāq̄
facit in terra cp in oppo
fito: pli.li.viij. cute veftes erpoliunt̄.

The *Ortus sanitatis*—literally 'garden of health'—was published in 1491 in Mainz and is among the earliest books in the Museum's library. It consists of 1066 chapters, each headed by a woodcut, with a description of the form and medicinal properties of substances in use by physicians at the time of Henry VII. It deals with 530 plants, 164 land animals, 122 birds, 106 fishes, and 144 stones and minerals, each of which was thought to have curative, health-promoting, or poisonous properties. There is no single author, the information being taken from numerous earlier works and thus giving a marvellous insight into the kind of medical treatment employed by practitioners living between 200 BC and AD 1500. For two or three centuries this book provided the standard reference for the medical profession.

Medicine and pests

In its origins and right up until the present time, medicine has had particularly strong links with natural history. One might almost say that it evolved out of natural history, for the first gatherers of herbs for curative purposes already had a well-developed knowledge of those leaves, roots, seeds, and fruits that were edible or useful, as well as a means of distinguishing them from those that were not. Presumably some fairly obvious reactions to plants with purgative or emetic properties led on to the refining of haphazard experience to produce a whole medical repertoire. This entailed accurate identifications and a kind of classification, in other words the basic taxonomic techniques used for understanding all other parts of the natural world.

For botany, the herbal played a critical role. Dating back to the Greek physician Dioscorides in the first century AD, it was a book for identifying medicinal or otherwise useful herbs, with descriptions of their properties and uses. In the early part of the sixteenth century it was the German 'fathers' of botany—Brunfels, Bock, Fuchs, and Cordus—who set out to bring Dioscorides up to date and in the process they made botany into a discipline in its own right. They did this largely by discovering that the German plants they so carefully illustrated simply did not match with those of Dioscorides (whose plants were from the Near East). It was in this period too that medicinal and other plants came pouring into Europe from the 'Indies', often made known by the pioneer missionaries, especially the Jesuits. From the Caribbean came the first samples of maize, brought back by Columbus; from Peru came 'Jesuits' bark' or cinchona, which provided the quinine treatment of malarial fevers; from the New World also came tobacco and such herbs as the emetic *ipecacuanha*; from Ethiopia via Arabia came coffee; and so on. The list is enormous and it forced the early botanists to broaden their horizons and to take note of whole groups of plants quite unrepresented in Europe. Although medical or 'physick' gardens had been a regular feature of monastic life, there now developed in northern Italy from the 1540s the first botanic gardens used for teaching purposes. What is interesting is that medicine continued to act as a direct stimulus to botany for another three centuries or more and even now, long after the chemical-pharmaceutical industry has made most of the old herbal concoctions obsolete, medical interest can still return to plants and thereby encourage new studies on their classification, identification, and relationships.

Medicine also had its effects on early zoology and mineralogy, although never to the same extent that it did on botany. For example, various earths such as the Terra Sigillata or Lemnian earth from the island of Lemnos were used as antidotes against poisons and snake bites, while bezoars or organic concretions of ellagic acid found in the intestines of wild goats, of Madagascan lemurs, and (as calcium phosphate) of South American llamas, were believed effective not only against poisons but also against infectious diseases. Sir Hans Sloane's pharmaceutical cabinet has two drawers of over a hundred mineral specimens (including no less than 260 grams of Arsenicum Album or white arsenic, a rather dangerous item one would think for the physician to Queen Anne). Sloane seems to have had faith in his Terra Sigillata, but perhaps his bezoars were more for curiosity since by then it was considered that ordinary chalk was just as effective—and cheaper. However, the use of such drugs did not have a profound effect on zoology or mineralogy and certainly was in no way the spur that it had been to botany.

Human parasites, especially fleas, lice, bedbugs, and various worms, have been known since ancient times, but medicine's real stimulus to zoology, and especially to entomology, came relatively recently when it was discovered that certain diseases are caused by minute internal parasites which have come from outside the body, being transmitted by insects and other animals. The classic case was the proof at the end of the last century that anopheline mosquitoes transmit malaria, but numerous other animal vectors of disease were soon discovered. This led to a really critical need for accurate identifications of the mosquitoes and other intermediate hosts of disease organisms and a considerably better knowledge of their life histories, so that not just the disease but also the carrier of it could be controlled. The Museum has played an active role in many studies on the identification and classification of such species or groups of species, as for example those on the blackfly genus *Simulium*, members of which are carriers of the parasite causing onchocerciasis or 'river blindness'. Modern work, again often based on research at the Museum, has sometimes shown that the identification of the insect or other vector of the disease must be rechecked. Thus, the apparently well-known mosquito *Anopheles gambiae* now proves to be a complex of six nearly identical 'sibling' species. Since they transmit both malaria and elephantiasis in Africa, it is essential that methods of distinguishing the species be found, for each has slightly different habits, tolerance to insecticides, and so on. The superficial similarity between certain species or races is often very great, so that the classic methods of taxonomy using diagnostic characters visible under the microscope must now be supplemented by analysis of other, perhaps biochemical, features. Thus medicine continues to press taxonomy forward in the fight against disease.

Some natural (and some unnatural) dangers have been brought, via medicine, to the scrutiny of natural history, usually to identify the animal or plant responsible, or to learn something of its distribution, habits, and what related organisms may also be a hazard. Venomous snakes have troubled almost all civilizations and their correct identification is of obvious importance, as is also that of the various poisonous plants, especially when a rapid identification may indicate the treatment required. Sharks are still a danger in the coastal waters of countries like Australia and in this case a specialist knowledge of shark teeth is sometimes necessary in order to identify the species from the wounds it inflicts. There are in fact thousands of cases where the identification of a natural enemy and a knowledge of its habits has helped to prevent or lessen the danger. Some unnatural dangers, such as a supper party with the Borgias in Renaissance times, have been the result of a misuse of natural history, but poisons have also played a useful role in hunting societies, being used to coat arrow tips or to drug fishes so that they can be scooped out of rivers or lakes. In turn, the pharmaceutical industry has shown a keen interest in such 'native' poisons once the plant has been identified.

Problems of human health are rivalled by those caused by pests, whether of cultivated crops or of all those other elements of the environment that are exploited. Under natural conditions the majority of pests are usually adequately controlled by their own enemies; the balance may swing one way or the other, perhaps as a result of changes in climate, but the pest is just one of hundreds or thousands of different organisms in a complex but relatively stable web of food and feeders, prey and predators. With the intensive cultivation of single crops, however, the whole balance is upset, the predators of the pest being unable to keep pace with or perhaps being eliminated by the unnatural environment created. In such conditions, a single species—such as a bole weevil, bud worm, or stalk borer—can become rampant. Here again, natural history makes its contribution, for the pest must be very carefully identified and information given on its relationships and habits. Possibly, its nearest relative is an innocuous species whose numbers are controlled by a particular predator. It may thus be worth introducing that predator and trying to control the pest in a biological way. Work of this sort is carried out by specialized laboratories or field stations, but their research must eventually rest on proper identifications of all the organisms involved. More than this, a taxonomist at the Museum can also sort out the literature on the pest, since it may have been referred to under several different names, each with a description of some facet of its biology and distribution. Putting these together, one may find that a considerable amount of information actually exists. Conversely, one can eliminate misleading references to species once thought to be the same. For example, an aphid on blackberries is of no consequence, whereas one of almost identical appearance is a pest on raspberries, so that it is no use eliminating brambles as a control measure.

To pests of this sort must be added all those plants and animals that interfere with human artefacts and activities, from clothes moths to midges on a summer evening. A review of the 'Special Investigations and Advisory Services' in the Museum's triennial *Reports* shows that there is always something new, often surprising. Snails may infest airport runways, causing aeroplanes to skid; is there a natural predator? Is damage to jet engines caused by a particular species of bird and what are its habits? What are the lizards that have suddenly invaded a Nigerian meat-canning factory? What is it that can eat holes in polyethene sheeting used for lining ponds and reservoirs? Can soil particles on the roots of the Museum's older herbarium specimens provide information on past levels of lead and trace elements?

It is this constant stream of inquiries that serves to break down the distinction between pure and applied taxonomy—to the mutual benefit of each. After all, the point of taxonomy is to use it.

The Scorpiones or scorpions, which belong to the Arachnida (spiders, harvestmen, mites), are an ancient group that has survived with little change since the Silurian period, about 400 million years ago. Their characteristic slender 'tail' ends in a sting which produces a venomous secretion. Contrary to popular opinion, the nature and potency of the venom varies considerably between species and in most is quite innocuous or produces only local reaction. However, some species of the family Buthidae inject a powerful neurotoxic venom, which can be fatal, particularly in young children. High fatality rates from scorpion stings have been reported from Mexico, the Durango scorpion (*Centruroides suffusus*) being the most dangerous. Shown here is the original specimen or type on which Reginald Innes Pocock, an Assistant in the Museum from 1885, based the description and scientific name of the Durango.

The ticks and mites (Acari) are a subclass of the Arachnida. Ticks are external parasites of land vertebrates and, with very few exceptions, all developmental (young) stages as well as the adults feed on the blood and tissue fluids of their host. While the irritation and discomfort caused by their bites are far from negligible, ticks are important to veterinary and human medicine because they can transmit disease organisms, including viruses, bacteria, rickettsiae, and spirochaetes. Shown here is the Rocky Mountain wood tick (*Dermacentor andersoni*), widely distributed in the western parts of North America and the vector of Rocky Mountain spotted fever (a severe typhus-like disease), as well as the virus of Colorado tick fever. It can also cause severe paralysis in humans and in cattle. The specimen is from Professor G. H. F. Nuttall's then unrivalled collection, presented to the Museum after his death in 1937.

Fleas play an important role in the transmission of diseases, since the adults of most species commute between various and often unrelated animals. There are more than 2000 different species and subspecies of fleas, the vast majority being parasites on land mammals, but about 6 per cent on birds. As carriers of plague (primarily infecting rodents, but also man) as well as other diseases, fleas are of great medical importance. The best-known transmitter of plague is the tropical rat flea (*Xenopsylla cheopis*), shown here in a superb drawing by one of the greatest of entomological artists, Amadeo Terzi (1872–1956). After participating in Manson's classic experiment at Ostia in Italy, where the transmission of malaria by mosquitoes was finally proved, Terzi moved to London and from 1902 spent the rest of his working life as an illustrator at the Museum; he contributed to over fifty books and some 500 scientific papers, claiming to have produced over 37 000 drawings.

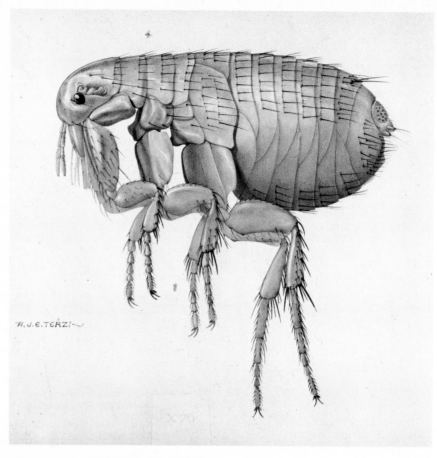

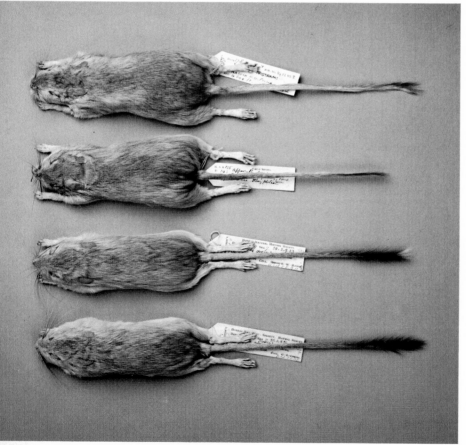

Plague is usually transmitted to people by fleas from house rats, but in fact many species of rodents can act as reservoirs of the disease. The susceptibility to plague, however, can vary greatly even between apparently closely related species of rodents. In the gerbils from Iran shown here, the upper two are of a species resistant to plague (*Meriones persicus*), while the other two (*M. tristrami*) are susceptible. To clarify the difference between such similar species, and thus place biological studies on a firm footing, the large collections of skins and skulls in the Museum are essential. The Black Death or bubonic plague, that spread from China to devastate Europe in the fourteenth century and remained endemic in London for 300 years, was transmitted principally by fleas on the black rat (*Rattus rattus*)—a species that occurs in a bewildering array of races and varieties, few of them black. Plague remains a serious disease in many parts of the world.

Schistosomiasis is a disease affecting some 200 million people in tropical countries; an alternative name for it is bilharziasis. It is caused by blood flukes of the genus *Schistosoma*, parasitic flat worms that live in the blood vessels surrounding the intestine or bladder. If the eggs, on being voided with the faeces or urine, pass into freshwater the second phase in their life-cycle takes place in certain species of snail. Finally, the small, free-swimming stage known as *cercariae* are released into the water and these enter the human host again through the skin of people washing or bathing. Some of the African host snails are shown here, of the planorbid genera *Bulinus* and *Biomphalaria*. Research in the Museum has been directed toward a better understanding of the biology of these snails and the worms themselves, with particular emphasis on their host–parasite interactions.

Locusts have been a scourge of crops ever since the development of settled agriculture over 10 000 years ago. Locusts compete with the farmer for survival in the warmer parts of the world where dependence on farming is most acute; the eighth plague of ancient Egypt was of locusts. Under certain conditions locusts can form migrating swarms; in the desert locust (*Schistocerca gregaria*—shown here) such swarms can contain as many as 40 000 million individuals, covering about 1000 square kilometres and consuming up to 80 000 tonnes of food a day (approximately the entire daily food consumption of the British Isles). In their non-swarming and solitary phase they live as harmless grasshoppers.

This wood wasp, *Sirex noctilio*, was accidentally introduced into Australia about twenty years ago and began to cause serious damage to plantations of softwood trees, especially *Pinus radiata* in Tasmania and Victoria. The females, which have stout ovipositors used for drilling through the bark and depositing the eggs, also inject the symbiotic fungus *Amylostereum areolatum* and a toxic mucus into the wood. The fungus provides food for the developing larvae, but also kills the tree. Biological control has been attempted, using various insect species that will feed on the eggs, larvae, or pupae.

The most promising have been the parasitic wasps *Megarhyssa nortoni* and *Ibalia leucospoides*. Nematode worms have also shown promise since they invade the eggs while still in the female, with the result that only eggshells filled with nematodes are deposited in the tree. The biological control of insects was first proposed by Erasmus Darwin, the grandfather of Charles Darwin, who suggested in 1800 the control of aphids by the larvae of hoverflies 'as the serpent of Moses devoured those of the magicians'.

Non-venomous snakes usually have long, pointed, and recurved teeth with which their animal food is caught, held, and by movements of the jaws drawn into the mouth and down the throat. The teeth of poisonous snakes are more specialized. In some, the fangs have a simple groove down which the venom travels and these lie at the back of the upper jaw. In the cobra and its allies, such as the green mamba (*Dendroaspis viridis*—shown here), the fangs are hollow and at the front of the upper jaw, thus resembling a hypodermic needle. In the vipers and distantly related mole 'vipers', the fangs can be folded away when not in use. It was not until 1648 that the Dutch physician Willem Pies, after working in Brazil, realized that the poison was injected in some species by means of hollow fangs. The staff at the Museum are sometimes consulted on the correct identification of venomous snakes, both to advise on dangerous species and in deciding on appropriate treatment for their bites.

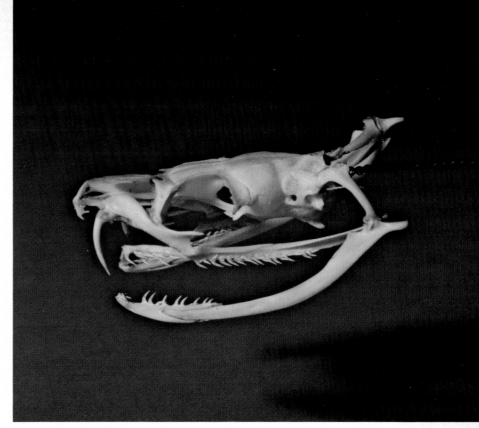

Some European and African malaria-carrying mosquitoes are in fact members of species complexes composed of 'sibling' species which can be virtually inseparable even microscopically as adults, but which have quite different habits and disease carrying abilities. Thus, harmless animal-biting species may strongly resemble harmful man-biting species that carry human diseases. Shown here is *Anopheles gambiae*, a most dangerous transmitter of both malaria and elephantiasis in Africa. Genetic research shows that there are actually six species, known now as the '*A. gambiae* complex of sibling species', each with distinctive habits, distribution, and medical importance. The Museum holds the types or original specimens of five of these sibling species and these have been of crucial importance in settling the names of the species and thus their identity in the field and in the literature. Other similar species complexes of mosquitoes and blackflies are well represented in the Museum collections and such specimens are invaluable in solving similar taxonomic problems in medical entomology.

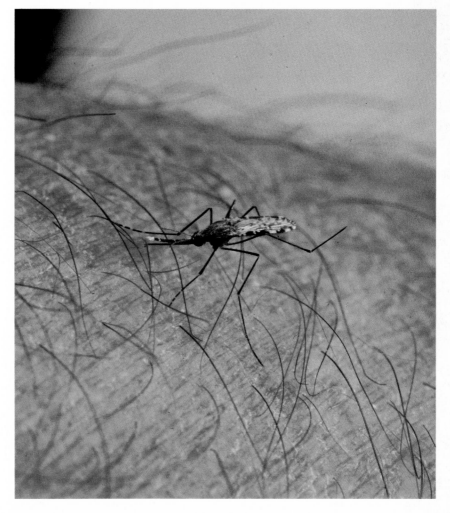

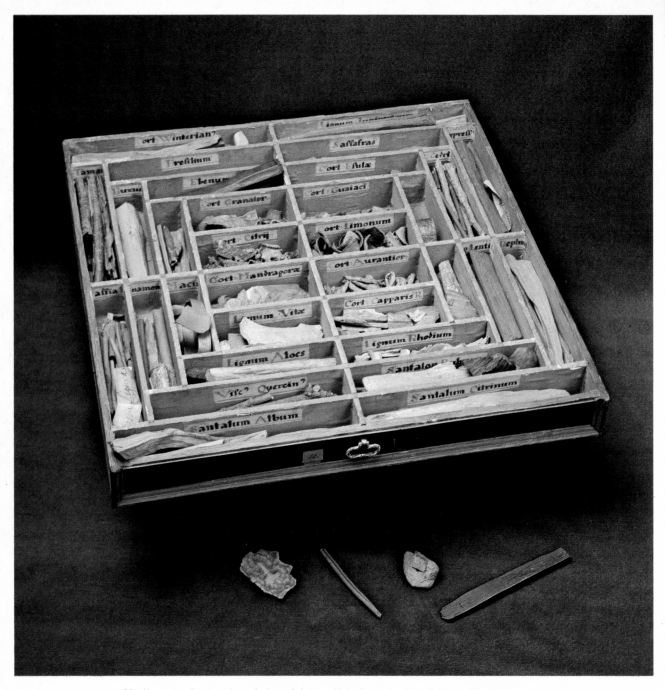

Until recent times, a knowledge of the medicinal properties of plants (and some animals) was an essential part of the physician's training. In London, such teaching was given by the Demonstrator employed by the Society of Apothecaries at their Chelsea Physic Garden, where new drug plants from abroad were grown. Interest in foreign drugs led to the collection and study of herbarium specimens and drug samples. In the possession of Sir Hans Sloane was this set of Indian drugs sent to Sloane by Dr Patrick Adair, an officer of the East India Company. Sloane was extremely careful with drugs whose effects seemed to be doubtful, but he extended the use of Peruvian bark (quinine) beyond that of fevers to the treatment of gangrenes, nervous complaints, and haemorrhages. He was the first to discover the nutritional value of a mixture of milk and chocolate and had this mixture manufactured and sold for consumptive and other cases early in the eighteenth century. Although many drugs are now produced synthetically, identification of the original plant sources can still require help from the staff of the Museum.

Decorative seeds have been widely used for personal ornament and for ritual purposes. Trinkets made from such seeds are nowadays often purchased by tourists visiting tropical areas and although most such seeds are perfectly harmless there are some which are highly toxic and are particularly dangerous to children. Shown here is a Zimbabwean specimen of the widespread tropical plant *Abrus precatorius*. Its red and black seeds are so poisonous that if even one were chewed and swallowed, it could prove fatal to a mature man. As a consequence, the staff at the Museum are frequently asked to identify seeds imported as beads and this work is greatly aided by the extensive collection of herbarium specimens and its supporting collection of dried seeds and fruits.

This is a reasonable match for the lectotype of *Cannabis sativa* L. in *Hortus Siccus* Cliffortianus and is probably of the same N.European stock

Cannabis sativa L. (s.str.)
Specimen grown in England in late 18th century
Determinavit W.T.Stearn

The name cannabis, like that of hemp, is of ancient origin and in fact both are ultimately derived from the same Asiatic word. The scientific name *Cannabis sativa* was first adopted by Carl Linnaeus in his *Species plantarum* of 1753 and again in the fifth edition of his *Genera plantarum* of 1754. For legal and other purposes it is critical to determine precisely the plant intended by Linnaeus, since the name has been used for a number of variants developed over three thousand years for fibre, seeds, and narcotic resin. Careful study of Linnaeus's writings and material shows that his *C. sativa* was based primarily on the North European race, long grown for its fibre (hemp) but meagre in narcotic properties. Cannabis plants are normally either male or female, and since the major taxonomic characters come from the fruiting material, a female specimen which Linnaeus himself had once examined was designated the type of *C. sativa*; henceforth, this specimen defines the meaning of the name *Cannabis sativa* for legal, botanical, or other purposes. This well illustrates the importance even today of some of the very old historical specimens in the Museum.

The fact that insects like mosquitoes and fleas can transmit diseases to humans and to livestock has been known since the turn of the century, but only recently has it been discovered that certain moths may also be vectors of disease organisms. The moth shown here, a species of *Filodes* from southeast Asia, normally feeds on the liquid bathing the eyes of cattle. By moving from one animal to another, it can transmit eye infections from diseased to healthy cows. Since it can easily be brushed away, it is not a serious human pest. The proboscis in most moths and butterflies is a delicate sucking apparatus but in the noctuid moth known as the vampire moth (*Calyptra eustrigata*) it is saw-like and is used to puncture the skin of mammals, especially cattle and deer, and to suck the blood from the wound. The vampire moth is therefore a potential transmitter of disease.

The long hairs of certain moth caterpillars are known to cause skin irritations. Far more dangerous are the secretions of the caterpillars of the South American emperor moth *Lonomia achelous*. It is the only moth known to secrete a chemical substance that induces bleeding of the mucous membranes in humans; untreated, it can last for over a month. It was first reported in 1967 and specimens of the caterpillar were sent to the Museum for identification. The patients had apparently brushed against a colony of caterpillars, which live on tree trunks and are well camouflaged. Local irritation of the skin was followed after a few hours by bleeding from the nose, ears, gut, vagina, and skin. It was subsequently discovered that the level of fibrinogen (responsible for clotting in the blood) was much lower than normal. By pinpointing the culprit, related species can now be tested in case they too are dangerous.

Insects are our main competitors for food from the land, and the list of such species is legion; they attack in the fields, in storage areas, in shops, and even in our homes. Among the least conspicuous are the stalk-boring caterpillars of certain moths, which tunnel into the stems of maize, sorghum, sugar cane, and other crops. As a result of their activities, there is usually a dramatic reduction in yield even if the plant itself still appears reasonably healthy. The caterpillar of the European corn-borer (*Ostrinia nubilalis*) is a major pest of corn and other crops in both Europe and North America,

causing losses of a million or more pounds annually. The Pyralidae, to which this species belongs, and the Noctuidae are the principal moth families with stalk-boring caterpillars. It was formerly thought that only one or two of these pyralid moths were involved. When it was realized that some 20 species were responsible, the problems of control measures in different continents were investigated properly. Shown here are *Ostrinia nubilalis* (*centre*) and the rather similar *O. furnacalis* (*left*) and *O. zealis* (*right*).

Blackflies of the genus *Simulium* occur everywhere and are usually harmless, but in tropical Africa and parts of the Americas they transmit the microscopic worms responsible for the disease called onchocerciasis. The worst effect, especially in West African savannah regions (where the disease is often called 'river blindness') is irreversible blindness on a major scale. Thirty years ago the Museum was asked to prepare a monograph on the *Simulium* of the Afrotropical region and recently a new monograph on the West African species has been initiated; this is intended to help with field identification in the huge Onchocerciasis Control Programme centred on the Volta River basin. Careful taxonomic work has shown that the 'species' illustrated here, *Simulium damnosum*, the chief transmitter of African onchocerciasis, is in fact a member of a complex of closely related (sibling) species; a similar situation occurs in *S. amazonicum* of the Amazon basin. The very large numbers of specimens sent from the Volta scheme for identification are invaluable additions to the Museum's collections.

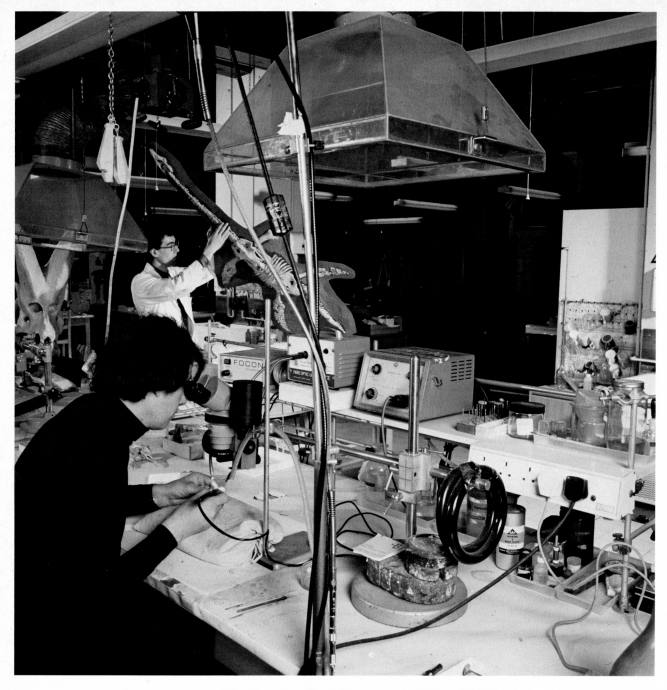

From the opening of the Museum at South Kensington in 1881 until 1926 the
laboratory for palaeontology was a two-roomed workshop where stonemasons
chipped away at fossils, and plaster casts were made. The methods were crude and
it was not until after the last war, with the development of new materials
(especially plastics) and new techniques, that finely detailed preparations of
specimens could be achieved. Engraving tools such as the dental mallet, driven
by air or by an electric motor and with hardened steel or tungsten-carbide points,
have replaced the stonemason's mallet and chisel, both for the delicacy they can
achieve and their speed. A binocular microscope is essential with the smaller
fossils, where fine engraving points like the dental ultrasonic dry probes may be
used; the miniaturized sand blaster or Airbrasive has also proved of value. The
modern laboratory is light and spacious and can accommodate any fossil,
whatever its size or state.

Techniques

Like any other scientific discipline, taxonomy or the theory and practice of classifying organisms has its own methods and technical 'props'. The most essential element, of course, is a large collection, since to classify is to compare and to draw conclusions. The specimens must be stored properly, with every effort to retain as much of their natural properties as possible. In using the collection, the amount of information that can be derived from it depends not only on the skill of the researcher, but also on the technical means available to explore in perhaps new ways, or to make traditional but extremely tedious examinations feasible for more than just the few specimens that sufficed for the earlier taxonomists. Equally important as a research tool is a first-class library. Finally, the data from such studies must be assembled to yield a classification and means of identifying future material, as well as perhaps providing some interesting biological, evolutionary, or other ideas. In this sphere computers are making a contribution, nor should it be forgotten that new theories for analysing biological or other data are as much techniques as the physical means for examining the specimens.

As far as the fundamental tool of taxonomy is concerned—the collections—there can be little doubt that the research workers at the Museum are among the most favoured. Perhaps more important than the fact that the Museum has about 50 million specimens, however, is the influx of specimens—sometimes nearly half a million a year. A considerable amount of that influx is material specifically requested from field workers or from other institutions, or else actively collected in the field by staff of the Museum. Such material is usually of direct interest to a particular study, with the result that when the work is published, the critical collection will often be that at the Museum, where it may attract further work and thus further specimens. To maintain such ever-expanding collections (for no specimen ever becomes redundant) and yet to be able to retrieve a particular specimen quickly requires considerable ingenuity in arrangement and cataloguing. A specimen within the Museum can often be located within a few minutes and in many cases lists and indexes produced by computer can give information on material from a chosen locality, collector, and so on. The older specimens, often with insufficient data on their labels or in the catalogue, pose serious problems and yet they are sometimes critical to the solution of a taxonomic question since they may be the original material on which a valid name was based. Thus, a taxonomist must have a good knowledge of the history as well as the contents of his collections.

Some of the preservation techniques used in the Museum have remained unaltered for perhaps two centuries, not because the Museum has lagged behind in modern technology but because of the risk involved. Alcohol has been used for the storage of many of the vertebrates, in particular the fishes, amphibians, and reptiles, and some of the specimens have remained in good or fair condition for a period of more than two centuries (their present state perhaps depending on how well they were 'fixed' initially). What guarantee is there that some of the modern preservatives will work for so long? A special laboratory exists for work on preservation and the preparation of tissues for particular types of study.

This was in fact the first European museum to develop an apparatus capable of freeze-drying entire biological specimens. For centuries, zoological and botanical specimens have been preserved dry, but with their true form distorted by shrinkage, whereas in the freeze-drying technique such distortion is avoided. The material is frozen at -10 to $-20°C$, then placed in a vacuum tank with a condenser whose coils are at $-40°C$ and the air pressure reduced to almost zero. The ice crystals in the tissue sublimate and the vapour recrystallizes on the condenser coils, a process continued until the specimen ceases to lose weight and is thus dry. Although not of great application yet for most taxonomic work, it is used for planktonic invertebrates, for archaeological and ethnographic specimens, for ocean bottom cores, for soft tissues to be viewed with the electron microscope, and for preparing material for exhibition.

A taxonomist examines and compares, continually searching for features which will relate two species or groups of species (genera, families, etc.) more closely to each other than to all other such species or groups. Every part of an animal or plant must in some way reflect its relationships, so that where superficial features seem to be similar the taxonomist is encouraged (often literally) to 'dig deeper'. There are many ways in which this can be done. New 'taxonomic characters' may emerge from simple dissection, for the earlier taxonomists were often content to examine just the outside of a specimen. In vertebrate animals the skeleton is full of interesting information, but the older method of boiling it down and removing the soft tissue is time consuming and not very efficient for small specimens. Current methods include the use of *Dermestes* beetles to strip off the flesh (especially good for bird specimens) and the staining of the bones red with alizarin dye while the soft parts are rendered transparent (of great use for small fishes, amphibians, and reptiles). A similar technique can be employed for cartilage and nerves, with the advantage that the organs or systems can be seen in three dimensions rather than as a thin section on a microscope slide. For specimens that cannot be dissected because of their rarity, or for which the gross shape of a bone or a simple count of the vertebrae is sufficient, radiography answers the purpose.

New tools suggest new avenues to explore. The first microscopists of the latter part of the seventeenth century discovered a whole new world which, with the refining of the microscope, not only revealed new animals and plants to be classified, but new methods of doing so. Even in large organisms, a taxonomic character could now be something quite minute, such as a distinctive banding on a muscle fibre or the ornamentation on a pollen grain. The invention of the electron microscope opened up yet another new world; the light microscope can magnify about 1000 times, whereas the electron microscope can magnify up to 1 000 000 times. The Museum now has equipment for viewing both thin sections and also for looking at opaque specimens so that the structure can be seen in three dimensions, with the advantage also that the depth of focus is about 50 times as great as that with the normal optical microscope. All this does not of course make the older methods obsolete, nor does it produce some kind of final answer; an animal or plant is made up of parts which in their intricacy may be minute, but in their organization and their clues to relationships may be as obvious as the posture of an elephant or the climbing habit of a vine.

For taxonomists, to 'dig deeper' is often to explore quite new features. Two apparently identical crickets may be found to produce quite different sounds which can be analysed using an oscilloscope; the exact composition of proteins can be determined in groups of species and the relationships suggested by this can be compared with those derived from other taxonomic methods; the number and shape of the chromosomes or thread-like bodies in the nucleus of the cell (which bear the genetic material) can also throw considerable light on affinities between species. For the palaeontologist a major problem is to find ways of getting the most out of a fossil, which may mean dissolving the matrix away to reveal the specimen in an unprecedented degree of perfection, or using a variety of specialized tools which have now replaced the crude hammer and chisel methods of the last century. Even more sophisticated equipment is available for the mineralogist, either for chemical analysis or for determining the atomic arrangement and the physical properties of minerals. For the taxonomist, new techniques—often derived from other spheres of science—are continually offering potential help in solving his problems.

Taxonomy produces data, often in large amounts. The computer now provides new ways of processing data, of recording it, and of making it available to others. In the case of information associated with the collections, such as localities, altitudes, habitats, dates of collection, or names of collectors, a computer can, from a single entry of data, produce a great variety of lists and indexes which can save the curator much tedious searching. Computers also assist in the process of classification by allowing large numbers of measurements or other observations on many individual specimens to be analysed rapidly, often showing up relationships which would otherwise have been unsuspected. Another application is that of computers programmed to identify specimens by feeding in their salient characters, thus saving the specialist from much routine work.

Fine collections and new techniques are nothing without sound theories to interpret the results. A century ago, evolutionary ideas were beginning to revolutionize taxonomic thinking; all organisms have past as well as present relationships. No such wide-embracing principle has since been discovered, although the development of genetics, of theories of animal behaviour, and above all of new insights into ecological relationships have had profound effects on taxonomy. No museum can claim a monopoly on new ideas, but much that is happening at South Kensington in both the theoretical and practical spheres has enormous relevance for workers elsewhere.

In 1964 the remains of a woolly mammoth and a straight-tusked elephant were uncovered in a clay pit of the Tunnel Portland Cement Company near Aveley in Essex. To bring these back to the Museum in their original state so that they could be studied and preserved demanded considerable ingenuity. The bones were first exposed, hardened, and photographed, and then enclosed in a plaster cocoon. First a layer of damp tissue was applied, followed by layers of hessian soaked in liquid plaster of Paris. The cocooned blocks were then transported safely to the Museum, where they were not only available for research but served for a very striking exhibit. The retrieval of such specimens, from building or other sites, must often be done rapidly and at short notice, perhaps at a time when weather conditions are far from ideal.

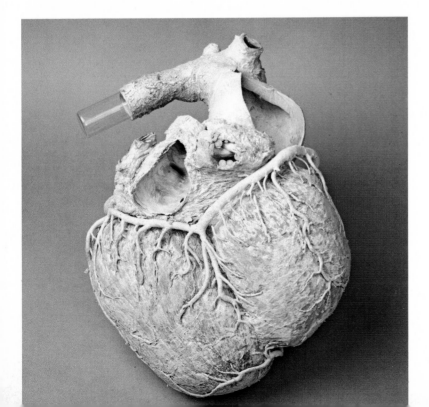

Freeze-drying, a technique pioneered by the food and pharmaceutical industries, was taken up by the Museum in 1964 and it was the first museum in Europe to develop an apparatus for the treatment of entire biological specimens. Equally important, it was shown that preserved material could also be freeze-dried, a technique hitherto presumed not possible because of the difficulty of diffusing vapour from ice crystals through preserved tissues. This massive elephant heart (5.9 kilograms), originally preserved in formalin, was successfully freeze-dried in 1970 and the method has become a routine procedure for preserved material. Freeze-drying is now used for many aspects of biological science and also in archaeological and ethnographic studies. In recent years it has been found a useful technique for the preparation of soft tissues for the scanning electron microscope.

The conventional light microscope, using special lenses and an ultraviolet light source, can magnify about 1000 times with a resolution of 100 nannometres (a 10 000th of a millimetre). The modern scanning electron microscopes can magnify as much as 1 000 000 times and have a resolving power down to 6 nannometres. This enormous increase in the detail that can be made visible has had profound effects on the study of pollen, diatoms, minute insects, minerals, small parts of organisms, and much else that was hardly guessed at previously. In 1966 the Museum acquired one of the first commercially available scanning electron microscopes; it now has two more modern intruments that are fully utilized by the Museum's scientists. By bombarding the specimen with a stream of electrons, a secondary stream is produced which is captured in a high voltage field, amplified, and used to modulate a cathode ray tube or television screen. This can be photographed as well as viewed and has the advantage that the depth of focus is about 50 times greater than that with an optical microscope.

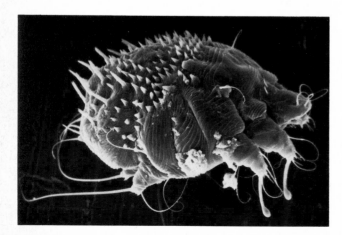

The scabies or itch mite (*Sarcoptes scabiei*) is an attractive subject for the scanning electron microscope. This mite, approximately 0.4 millimetres long, is a skin parasite of a wide variety of mammals, including humans. Mites from one host species will not normally establish themselves permanently on another species, but it has not yet been possible to distinguish between the forms that occur on different mammalian hosts. Having found a suitable host, the mite burrows into the skin—the hands, feet, and scrotal area being particularly favoured. Infestation causes severe itching, probably because of excreta in the burrow of the mite, and the resulting scratch marks give the appearance of a rash. Even one or two adult *S. scabiei* are sufficient to produce the symptoms.

106

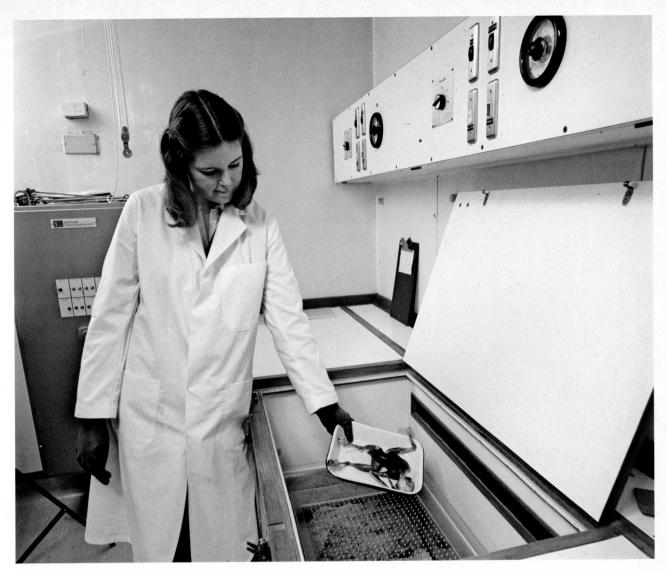

The skeletons of vertebrates are of great interest in research on the morphology and evolution of the various groups, but the production of good skeletons by boiling and manual dissection is time-consuming. For birds, the most convenient method is to use a South American beetle, *Dermestes punctatus*, closely related to the bacon beetle and the carpet beetle of Britain (both serious museum pests). The beetle colonies are kept in a specially designed building apart from the research collections and under controlled conditions, where they are encouraged to feed on bird specimens until all the flesh has been removed. If the resulting skeleton is held for a while in the smoke of burning magnesium wire, it turns a brilliant white, the sutures between the bones are enhanced and first-class photographs can be made. This method is suitable for any small vertebrate, but in fishes it is preferred to clear the flesh with enzymes, thus making it transparent, and then stain the bones brilliant red with alizarin, the specimen thereafter being kept in glycerine.

Although a fossil must necessarily give far less information than the complete animal or plant, it is surprising how much can be learned. For example, the musculature of a dinosaur can to a large extent be reconstructed and from this its posture, speed, and something of its habits can be deduced. Much depends, however, on the state of the fossil and the degree to which it can be prepared in the laboratory. One most useful technique, first tried successfully by the staff of the Palaeontology Department, is the use of a weak acid to dissolve away the matrix enclosing the fossil. Dilute solutions of acetic acid will dissolve carbonates, so that fossil bones can be completely freed from limestone. Formic acid can also be used or, for ironstone rocks, a solution of thioglycollic acid. At certain stages it may be necessary to use synthetic resins as hardeners (consolidants) and as adhesives. Shown here is the skeleton of a coelacanth prepared in this way.

The General Herbarium of the Museum contains the collections of flowering plants and conifers from all continents except Europe. Some 1.7 million sheets, on which the pressed specimens are mounted, are kept in 2500 wooden or metal cabinets and are arranged in a sequence of families generally following the classification of Bentham and Hooker. Nowadays, about 10 000 specimens are added each year, while some 7000 sheets are sent out on loan to other institutions annually. A large collection of dried fruits and seeds, as well as some 250 000 plant illustrations, are also kept in the Herbarium. The historical basis of the collection is the herbarium of Sir Hans Sloane, in 337 bound volumes, but perhaps the most celebrated individual item is the herbarium of George Clifford, the species in it having been named by Carl Linnaeus himself while he was under the patronage of Clifford in the 1730s. A herbarium, even the size of that at the Museum, is an efficient self-indexing data-bank and retrieval system, for in this room it takes less than three minutes to locate any one of the 150 000 species represented, or five minutes to locate any particular one of the 1.7 million individual specimens.

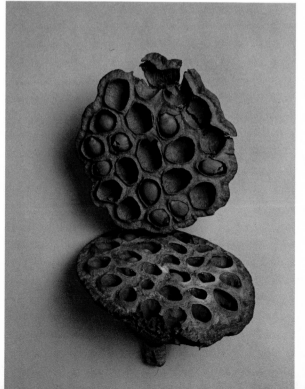

The older specimens in the collection sometimes produce unexpected information. During the last war, efforts to extinguish a fire bomb on the roof of the Museum resulted in a part of the Sloane collection of herbarium specimens being soaked. Not long after, it was noticed that certain *Nelumbium* seeds had germinated, showing the amazingly persistent viability of some plant seeds as compared with only a few days in others. In the middle of the last century Robert Brown succeeded in germinating seeds of the North American lotus *Nelumbium luteum* (shown here) from the same container in the Sloane collection, then more than a century old. This was repeated more recently by the late John Ramsbottom at the Museum, the seed producing a 1-centimetre shoot within 24 hours although the seed was over 200 years old. Equally unexpected was the discovery that minute soil particles adhering to the roots of old herbarium specimens could provide precious information on the history of now altered soils.

For the size of its collections, the Museum is perhaps unrivalled, and in certain groups of animals, plants, and minerals it certainly has the most comprehensive material available for study and comparison. Although large numbers of specimens are loaned for study to other institutions, their scientists try to pay at least one visit (but often many more) to use the collections on the spot. Such visitors range from the taxonomist en route for Africa or the Far East who stops for a day to check on some puzzling point, to the student working for a Ph.D., who may spend three years or more patiently revising the classification of a group. In terms of visitor-days, the Museum is host to about thirty thousand each year and its contribution in this sphere can be seen in the numerous research papers or books that acknowledge its help. Among the visitors are also members of the public seeking information, either from the research departments or from the libraries, artists illustrating books or stamps, authors checking details, amateurs with material to identify, and a host of others for whom the Museum serves as a kind of encyclopedia for natural history.

Fossils often differ in colour only slightly from the matrix in which they are embedded and the details are correspondingly difficult to see. For this specimen of the palaeoniscid fish *Amblypterus*, the Museum's photographers devised a special technique, using an infra-red false-colour film. Normal colour films have three emulsion layers, sensitive to red, green, and blue light, but in this case the layers are sensitive to green, red, and infra-red, producing respectively magenta, yellow, and cyan, the result depending on how strongly the infrared is absorbed or reflected. Although the colours are thus distorted, the effect is to differentiate between visually similar minerals by artificially translating the infra-red spectrum into different colours.

Almost all the economically important ore minerals are opaque and thus cannot be studied in thin sections by transmitted light under the petrological microscope. When such sections are polished, however, they can be examined in reflected light; when this light is split up into its spectral components, i.e. colours, the degree to which particular parts of the spectrum are reflected can produce a characteristic pattern for each mineral. Such accurate reflectance values, which can 'finger-print' an opaque mineral, are known for very few ore minerals since measurement is difficult and requires complex apparatus. The Zeiss photometric microscope shown here is the most advanced of its kind. From its results a data-bank is being formed with the intention that, when sufficient minerals have been measured, a series of less complex criteria can be established so that identifications of these minerals can be undertaken elsewhere using much simpler microscope photometers.

The impact of computers on the biological sciences and in mineralogy has been profound. In September 1974 the Museum acquired a Varian computer and by the following year this was in use on various mathematical and statistical problems resulting from studies of morphology, mathematical ecology, and the use of numerical taxonomy in the classification of animals and plants. By June 1977, it was put to its alternative task of producing catalogues, and the two activities are now about equally shared. Statistical work has become increasingly important in museum studies now that large samples are used; the computer can not only save hours of calculations, but can also draw conclusions in no way obvious from a mass of raw data. For cataloguing, a program 'prompts' the user to enter data according to the rules. The computer can also be programmed to identify specimens, the operator merely entering the salient characters. Shown here is one of the terminals in use to identify a grass.

Physical anthropology, or the study of the bodily characteristics of human populations, their variation, and evolution, supplies valuable information not only on the origins and affinities of various ethnic groups, but also on the effects of intermixture and isolation between communities. Here a Faroe Islander, a member of a comparatively isolated population, is being measured, while notes are taken on his daughter's hair colour and other genetic characters. Such studies are a regular feature of research in the Museum. The data from these surveys, stored on standard sheets and on computer tape, adds to an existing data bank which is available for reference to other specialist workers. The Faroese are among the closest living descendants of the Vikings, but isolation and small population size resulted in an array of genetic characteristics distinctive among northern Europeans and perhaps no longer reflecting those of the ninth-century Norse colonizers of the North Atlantic islands.

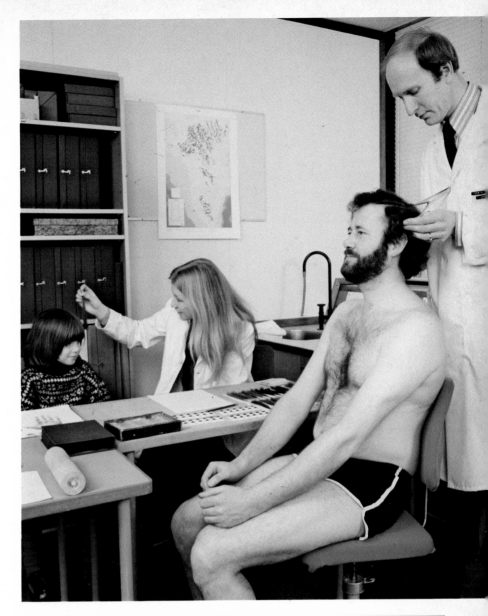

Perhaps the most impressive and best preserved of prehistoric human skulls is that of Rhodesian man. Since its discovery in 1921, interest has continued and there has been an increasing demand for casts of the skull. In former times, moulds were made with plaster of Paris and for something as complex as a skull it might require up to 40 pieces which had to be carefully removed and boxed together for the final plaster to be poured in (*right*). The method is still used, but with polyester resins and fibre glass as a mould. Far simpler are one- or two-piece moulds of silicone-rubber, which also allows for sharper definition (*left*). Plaster casts are heavy and easily broken; hard polyester resin is preferred nowadays, particularly for material that is frequently handled.

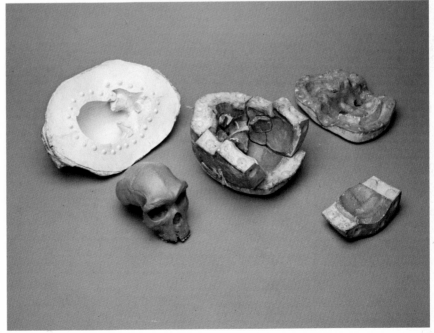

The unrivalled collections of natural history books, periodicals, maps, manuscripts, and drawings held in the Department of Library Services are essential for the success of research work carried out by the Museum's scientists and visiting research workers. Central to the work of the Museum is the classification, identification, and naming of animals, plants, fossils, rocks, and minerals, but without comprehensive collections of the relevant literature the taxonomist may needlessly repeat earlier research work, supply unnecessary new names, or fail to take account of earlier studies. For its nomenclature, taxonomy is the only discipline in which a search through the literature back to the mid-eighteenth century (or sometimes beyond) is obligatory. From an original five thousand volumes brought to South Kensington from the British Museum at Bloomsbury in 1881, has grown a library second to none for taxonomic natural history. It contains some 17 600 periodical titles (of which 8000 are currently published), about 750 000 volumes, 70 000 maps, and one of the finest collections of early natural history drawings and manuscripts.

In research on aphids (greenfly) in the Museum, the rearing of live insects is an essential method for understanding relationships between the species in this economically important group. Reared on leaves in tiny cages, the aphids can be subjected to various environmental conditions, which may have a profound effect on the shape, colour, and size of individuals. In this way it can be shown that variant individuals in the collection (which may have no environmental data recorded) are not distinct species, but have resulted from particular circumstances, such as an unusually warm summer. Genetic aspects can also be investigated, again leading to a better understanding of the species and their relationships.

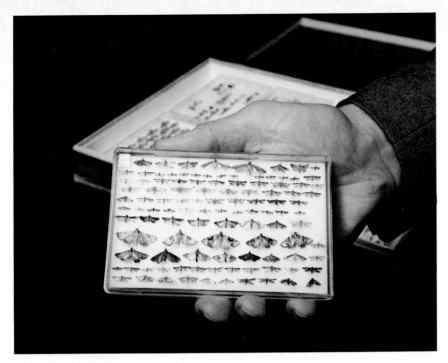

An essential aspect of field collecting is to bring the specimens safely back to the Museum. For butterflies and moths exceptional care is needed to prevent loss of the scales on the wings, most especially in the case of the Micro-lepidoptera (where some have a wingspan of less than 3 millimetres). These tiny moths are collected individually in small glass tubes and after killing are carefully pinned through the thorax with a minute pin and placed in a transparent box ($12 \times 8 \times 2$ centimetres) with a soft lining. Only at this stage are the wings carefully spread, using forceps or a fine pin. Once filled, the box is sealed with tape to prevent damage from pests. It occupies a minimum of space and can be easily inspected by customs officers or given a preliminary study by specialists. Shown here is a box of Microlepidoptera collected in 1976 in the Hawaiian Islands.

Photography has become an essential tool in the research carried out at the Museum, not only for the presentation of results but also for the recording of specimens, preparations, and dissections whether for future study, for lectures, or for archival purposes. In addition great demands are made on the Photographic Unit for exhibitions and for the Museum's publications. The well-equipped Unit must employ techniques which are varied, often highly specialized, and sometimes unique. Here a photographer is working some 30 metres above a quarry floor in the Mendips, recording the earliest evidence of man in the British Isles (Middle Pleistocene or about half a million years ago). A special seat, designed by the photographer himself, is fitted to a ladder and swung out from the face of the quarry, with safety harnesses in case of trouble. From this vantage point the photographer was able to record the bedding planes of the rocks, the resulting photographs being used by the scientific team to obtain accurate measurements and records.

Although the image of the Museum as a place essentially for children is rapidly
changing, it is on children that society's future attitudes to nature depend.
The Children's Centre, established in 1948 by a London teacher, is now run
entirely by the Museum's staff and since 1969 has occupied a large room in the
North Hall. Intended for children up to 15, it offers opportunities to handle
specimens, use microscopes, follow trails round the Museum, draw specimens, and
answer quizzes—or just browse. During holiday periods, there may be as many as
700 eager young naturalists to cope with daily. The room is strictly for children,
but during the longer school holidays a Family Centre is organized for both adults
and children. Saturday Clubs are also run for regular young visitors.

The Museum and the public

The notion of a museum as a place where ordinary people can come and learn took a long time to develop. It had its roots in the Renaissance cabinets of 'curiosities', but these were chiefly kept by the nobility or by scholars, to be shown off to equally noble or scholarly friends. The curiosities excited general wonder, with some attempt to fit them into the known order of things, but more often to assign to them marvellous properties. The museum formed by John Tradescant at South Lambeth in London had something of this flavour, for here one could see the hand of a mermaid, a feather 'from a phenix wing', the horn of the sea-unicorn, and much else. The museum was later opened to visitors—the first public museum in England, perhaps in Europe. It was well arranged, but in a sense natural history could not teach until it had learned.

By the mid-eighteenth century, things had changed. Objects still tended to be 'curiosities' and some retained their magical reputation, but sufficient order was being brought to the natural world that more often than not a new object had a place in the scheme. At the same time, the organization of this scheme of nature involved many more people than before, from those who studied it to those who collected for it or had a material interest in the uses to which the new minerals, plants, and animals could be put. The time was ripe, not only for the museum serving some particular learned society or institution, but for museums to be patronized by the public.

To Sir Hans Sloane must go the credit for envisaging a truly national museum. He himself was delighted to show his own collection (until he found specimens missing) and he had undoubtedly read Robert Hooke's dictum that a museum is a place 'where an enquirer ... might peruse, and turn over, and spell, the book of Nature'. In his will, Sloane specified that his collections should be available 'for satisfying the Desires of the Curious and for the Improvement of knowledge', to which the 1753 Act that brought the British Museum into existence added that the collections should be 'not only for the inspection and entertainment of the learned and curious, but for the general use and benefit of the public'.

Unfortunately, the early history of the British Museum is very much a lesson of what a public museum should *not* be. As much material as possible was placed on display, the rooms being crowded with cases of specimens. More serious was the fact that entrance was by ticket, which needed to be applied for beforehand, and tours were arranged in parties, with no time to 'peruse the book of Nature', let alone turn its fascinating pages. It was over a century before the Museum was open on all weekdays, and Sunday opening did not come until 1896. Unlike the private museums of the time, the British Museum was free (subsequent attempts to charge in 1923 and 1974 both failed), but it was far less popular than William Bullock's splendid Egyptian Hall in Piccadilly or Sir Ashton Lever's Holophusikon in Leicester Square.

By the 1880s, however, attitudes had changed and with the removal of the natural history collections to South Kensington it was possible to start completely afresh. Both Richard Owen and William Flower who succeeded him as Director of the new Museum were deeply interested in the role of museums in education, but their methods continued to reflect the needs of their age, in which classification was a dominant theme in biology. Galleries were apportioned to the Departments and within each gallery the systematic arrangement of the objects proceeded round the wall-cases. The book of nature could now be perused, and indeed it was, by amateurs and collectors who came to identify what they had seen, or by those who sought to be reminded of the vast wealth of the natural world. But there were many anomalies. Plants, for example, occupied a mere 3 per cent of the exhibition space according to the 1886 *Guide to the Museum*, invertebrates 18 per cent, but mammals no less than 41 per cent.

As William Flower put it, 'a museum is like a living organism—it requires constant and tender care.' By the 1930s there was a shift of emphasis in the displays, from pure classification of the objects to aspects of biology. In refurbishing the Fish Gallery around 1937, for example, over half the wall-cases were given over to aspects of feeding, breeding, camouflage, and so on. In doing this, the actual number of specimens was drastically reduced in favour of representatives of major groups, although the exhibits could still be used for reference to a large extent.

The modern era has produced even more fundamental changes, partly because of a shift in the emphasis of natural history and partly because of the revolution in schoolroom teaching (in turn based on new theories of learning). Detail, as is well known, tends to be forgotten rather quickly unless placed in some structured context. For example, rather than have cases full of real specimens it is thought better to take a few models and illustrate their relationships in some visually simple way, since the mere placing of numbers of specimens together is unlikely to produce any deep thoughts in the unimaginative child, while in the imaginative one it may lead to interesting but irrelevant ideas. By posing a set of questions and supplying the viewer with a choice of answers, the real significance of an exhibit will become apparent. This is quite different from an art museum, where the viewer is encouraged to frame his own questions. Again, the word 'display' is now seen to imply enjoyable window-gazing, whereas proper learning and retention demand a much more active involvement in an exhibit. In many of the modern exhibits the visitor is now invited to operate the model in some way, either physically to simulate the process or at least to involve the hands in the learning process. For small children, the sheer size of a large animal or its familiarity from story books will ensure that it remains a memorable experience. For older children, the demonstration of some biological principle, even one so simple as camouflage, will be thought about afterwards. The student, on the other hand, has come to learn and, although competent to manipulate abstract ideas, finds it very much easier if these are tied to a concrete reality, for which even a well-illustrated textbook is no substitute. Adults have a variety of motives, from trying to teach their own children to making a bee-line for some particular exhibit or topic of interest to them. To design exhibits suitable for all has become a science in itself.

Not only are the methods of exhibiting different nowadays, but the scope of natural history is seen in quite different terms. It is realized that no major natural history museum can now ignore the biology of human beings themselves, nor the energy-sharing relationships of communities of organisms—the field of ecology. Equally important, the fact of diversity in animals, plants, and minerals is less interesting nowadays than the reasons why it has occurred. The unity of the natural world; our own place in this unity; and the processes by which it has come about and operates and can be changed in the future—these are the aspects which the Museum attempts to illustrate in the public galleries.

The Museum's commitment to its public, however, does not stop short at exhibition. It is regarded as a sort of living encyclopedia of natural history and a steady stream of specimens flows in to be identified. Some are frankly trivial, many are interesting, and a few are of great importance. All of them, however, are a constant reminder to the research worker that among the aims of natural history is the need to promote and encourage an interest in all aspects of the natural world. The specimens, the letters of inquiry, the sketches, and the photographs also show the direction that public interest takes and not a few research studies have begun simply to satisfy some inquiry. Specialist groups also look to the Museum as a source of help and encouragement, whether it is an aquarist circle that requires a speaker, or a local entomological club with problems of literature. In return, the contribution of amateurs to the subject is often impressive.

Less directly, the Museum serves the public through the advice and help that it gives to numerous local, national, or international bodies, ranging from municipal health authorities that want the contaminant of their water supply identified, to such organizations as the Food and Agriculture Organization of the United Nations, the World Health Organization, and many more. Other bodies, such as the Commonwealth Institute of Entomology, are dependent on the Museum's collections and their specialists actually work within the Museum. In this way, the needs of a very large 'public' indeed come within the orbit of the Museum.

Our relationship with the natural world in the future—and this includes the sheer enjoyment of it too—is a responsibility for everyone. It depends on a clear knowledge of the environment and of our own role as users of it and as an essential part of it. For this reason, the Museum's research and educational activities are not so separate as they might appear. Both try to contribute to the understanding, management, and enjoyment of the animals, plants, and minerals around us.

No museum can afford to refurbish all its galleries simultaneously; each must take its turn. As a result, there is always a medley of the old, the new, and the in-between. The Entomology Gallery represents the in-between, having superseded a rather shabby collection of table-cases without any real theme, but pre-dating the modern trend toward fewer specimens but more explanation of biologically important principles. Completed in 1970, it was conceived as an educational exhibit for fifth- to sixth-form children, but with enough for people of all ages. No label is more than 2–4 lines long, familiar insects are stressed, and biology is included as often as possible. Insects are arranged systematically on the left of the gallery, topics on the right, and economic entomology explained in the centre. The cross-gallery at the entrance contains a complete set of the larger British butterflies and moths with their larvae, since it is to Lepidoptera that most young entomologists turn.

This water-colour by George Scharf (1788–1860), a German refugee from the Napoleonic Wars, is one of the few pictures of the natural history exhibits at Montagu House, first home of the British Museum; if contemporary visitors are to be believed then the collections were actually a 'strange "Mischmasch" of works of art, natural curiosities, books and models' (as Prince Pückler-Muskau expressed it in 1826). 'Nothing is in order, everything is out of place' complained another visitor. A critical shortage of staff was partly responsible, but improvements came with the opening of the present British Museum building on the same site, more or less completed by 1852. Only with the transfer of the natural history collections to South Kensington in 1881–3 was it possible to give their exhibition proper scope. Nowadays, however, with some three million visitors a year and with a vastly wider range of topics to present, the problems of space and adequate staff are felt again.

This elegant case of humming birds probably dates from the early nineteenth century and may well be that listed in the sale catalogue of the Egyptian Hall, being William Bullock's Museum at 22 Piccadilly, auctioned in 3342 lots in 1819. It is a particularly fine example of this kind of display, where a pseudo-natural 'prop' (in this case a highly contrived, lichen-covered tree) is used to set off the greatest number of objects (about 90 birds). Such small birds would have been lost in a more realistic diorama, while the sheer fun of exploring their diversity would be spoiled if they were pinned like butterflies to a painted background. Although this type of exhibit belongs to the Victorian era, it is very popular, partly because it is a beautiful thing in itself, and partly because visitors have to ask their own questions, such as the reasons behind such diversity. Science is not so much answering questions, as finding new questions to ask. After all, it was the diversity of finches on the Galapagos Islands that set the young Charles Darwin thinking.

One of the main aims of the more recent exhibitions in the Museum is to stimulate visitors to think actively about natural history. The exhibition 'Dinosaurs and their living relatives', for example, has two parts: the first being the specimens themselves, the second showing how scientists try to discover the relationships between dinosaurs and other animals, both living and extinct. Visitors are invited to follow the line of argument by looking at the displays and using the information in them to answer a series of questions.

Young visitors are here using 'activity sheets' in the ecology exhibition. A wide range of such sheets is available from the new Visitor Resources Centre, covering human biology, dinosaurs, human evolution, and other subjects. The Centre also assists teachers in planning visits with school parties, and helps visitors with leaflets and information on the galleries.

In the nineteenth-century museum it was the sheer diversity of animals, plants, and minerals, many of them recently discovered in newly explored lands, which constituted an exhibit. Naturalists were preoccupied with cataloguing and classifying them, a task still as important, but nowadays overshadowed by striking biological advances or such pressing social themes as conservation. The former Fish Gallery, dating from 1937, economized on the classificatory aspects by devoting half the wall space to biological topics, but there were still sufficient specimens for the angler to identify his catch or the enthusiast to locate a representative of most of the fish families. There were fine anatomical cabinets, and there was a full-sized cast of a whale shark, the largest fish known. This gallery, dismantled in 1976, marked an important step in the evolution of exhibition techniques in the Museum.

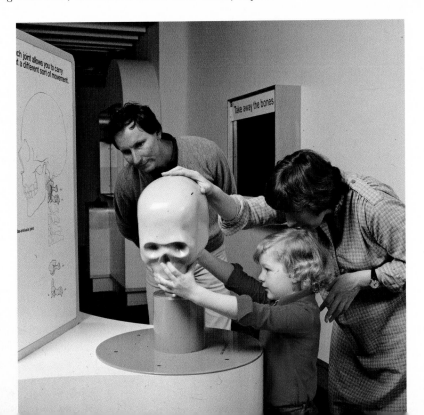

The Hall of Human Biology has the apt subtitle 'An exhibition of ourselves'. It has been designed to allow visitors to discover the functioning of the human body for themselves. Here, we see visitors in the 'movement' section exploring the relationship between movement and joints. Throughout the exhibition, visitors are encouraged to find out about themselves by interacting with the displays—not only models, but a variety of games and audiovisuals. Sections on fertilization, birth, growth, movement, the brain, hormones, learning, perception, and mental development are linked logically to give an over-all picture of how the body functions.

The exhibition galleries are the most obvious contribution that the Museum makes toward education and the dissemination of specialist knowledge. There are, however, many other contributions, such as those to amateur groups who require material to be identified, literature to be recommended—or sometimes just encouragement for their activities. For anglers, a critical identification may be necessary, as when a specimen of that most popular fish the roach (*Rutilus rutilus*) fails to beat the British record because examination shows it to be a hybrid with a chub or bream (both of which are larger species). The current British rod-caught record, a female of 4 lb 1 oz from a gravel pit near Nottingham in 1975, is shown here. The distinction between the roach and its various hybrids can best be made by examination of its pharyngeal or 'throat' teeth (see inset).

Large numbers of objects are brought every year to the Museum to be identified. They are sent on to the relevant Department and in most cases are bones, fossil shark teeth, curious stones, seeds, and the like picked up on holiday or found while digging in the garden. Of the more unusual was the object shown here, found in an Irish peat bog. X-ray analysis showed that it was composed of almost pure apatite or bone mineral and it was passed from one department to the next until finally identified as a large hydatid cyst or bony secretion formed in humans in response to the invasion of the body by a cestode worm. Some of the objects sent are afterwards donated to the Museum and may be of considerable interest; of hydatid cysts, the Museum then possessed none so large (14 centimetres in diameter).

From time to time the staff of the Museum are asked by health authorities to identify 'foreign bodies' found in various foods, which could lead to a serious complaint or even court action if the object is not a legitimate part of the food. It may be necessary to examine hair or bone by the scanning electron microscope, but the objects shown here can be identified by an experienced mammalogist using a hand lens. Individually they may resemble small talons from a bird or even the 'spurs' from a spiny dogfish, but they are in fact the quite normal keratinized papillae found on the hind part of the tongue of an ox. Their presence in a meat pie is therefore quite legitimate. On one occasion the identification of a small fish from a meat pie prevented serious repercussions: it was clearly a common aquarium fish and its uncooked condition led to the true explanation—a prank by the small son of the complainant.

This unique model, a reconstruction of a Neandertal woman based on accurate measurements from a skeleton some 41 000 years old found at Tabūn in Israel, forms part of the exhibition 'Man's place in evolution' (opened in 1980). The exhibition looks at the way human beings are related to other animals, both living and extinct, and explains that the Neandertals were our closest fossil relatives; they had brains as large as ours and they made tools, used fire, and had ceremonies for their dead. Modern exhibitions require many people and skills. In this case, a team of Museum scientists, designers, writers, and educationalists produced a plan, which was then passed to draughtsmen, illustrators, engineers, photographers, modelmakers, and taxidermists who, with the help of contractors engaged through the Department of the Environment, actually produced the exhibits. Here, the latter include casts of classic fossils, photographs, illustrations, an audiovisual programme, a computer-operated question-and-answer display, and a number of superb models.

The Museum's display of minerals is perhaps the most comprehensive in the world, offering to the student a visual aid impossible to rival by books, and to the non-specialist a feast of colours, shapes, and unexpected guises for familiar minerals. It is the only main gallery to remain in practically its original form, apart from the modern meteorite pavilion at the eastern end. In fact, the exhibition cases and even the locks on them are of that fine Victorian craftsmanship with which the Museum was built. The displays themselves, on the other hand, are constantly up-dated to keep pace with modern research and some extra cases have been installed to display large specimens more attractively.

Temporary exhibitions are a useful way of high-lighting topical problems in natural history or recent advances relevant to the work of the Museum. A continuing programme of temporary exhibitions is now mounted in the North Hall and has dealt with such subjects as *Jojoba*, a new desert crop plant whose seeds could provide a substitute for sperm-whale oil, and with the 'locust menace', which coincided with the beginnings of an ominous upswing in the world population of desert locusts (*Schistocerca gregaria*) early in 1979. Shown here is 'Patterns of diversity', a description of the work of the Royal Geographic Society's expedition to the Gunung Mulu National Park in Sarawak in 1977–8, on which some members of the Museum took part and others identified the material brought back.

The postcard counter has always been a traditional stop on any museum tour. For 50 years, from 1921, the Museum had a small stall on the right of the entrance, but in recent years more extensive shops have been developed to cope with the growing numbers of visitors and the increasing variety of publications produced by the Museum. These range from the traditional postcards to colourful guides, highly illustrated manuals for the identification of fossils, and comprehensive monographs and reference works for the specialist. Accurate models and casts are available along with many other gifts, souvenirs, and books, all selected with natural history and conservation in mind. Meanwhile the small shop at Tring was enlarged and entirely refurnished by the Museum's craftsmen using redundant timber from cabinets and shelves installed at South Kensington in the 1880s.

From the early catalogues of the contents of the British Museum gradually evolved the research papers that from 1950 were published as *Bulletins* of the Museum. The early guides to the Museum have shown a similar development and to them have been added books, pamphlets, leaflets, and wall-charts which enlarge the scope of the exhibits. In 1969 came a new departure, the facsimile reproduction of examples from some of the famous collections of natural history drawings owned by the Museum, in this case water-colours of fishes executed on Captain Cook's voyages by Sydney Parkinson, George Forster, and others. This was followed by a selection of superb Chinese water-colours (of animals and plants) from the collection made by John Reeves at Canton. The quality of the reproductions, printed by the old-fashioned collotype process, has perhaps never been surpassed.

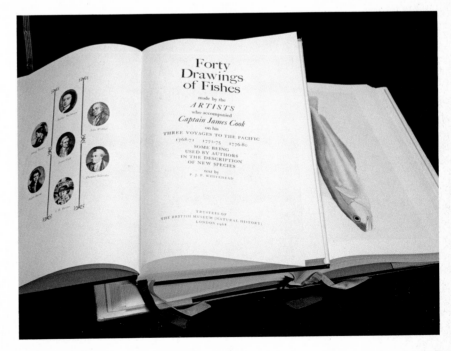

From the Zoological Club formed in 1822 by members of the Linnean Society developed the Zoological Society of London, which in April 1828 opened to the public the now famous London Zoo, the first to be built in Britain. Its relationship with the Museum has been a close one and the Museum has frequently benefited from specimens for both research and display. Among the most popular exhibits in the Museum is Chi Chi, a Giant panda (*Ailuropoda melanoleuca*) caught in the mountains of Szechwan in China on 4 July 1957 and, after a year in the Peking Zoo and exhibition in various countries, bought by the London Zoo. Unsuccessful attempts were made to mate her with An-An in Moscow in 1966 and again in London in 1968. She died of old age on 22 July 1972 but will now be remembered by a generation of children too young to have seen her alive.

Taxidermy, or the preservation and arrangement of animal skins in a life-like manner, demands as much artistic as scientific skill. A thorough understanding of the anatomy of skeletons and muscles, and the way that these interact, is essential, as also is a knowledge of skin technology. At the same time, the taxidermist must have an eye for detail and a training in various sculpting techniques with wood and metals. The standard method is to remove the skin and preserve it with a mixture of salt and alum. The desired posture of the animal is then constructed, using wood or metal for support, and to this an exact replica of the body is added, for which wood-wool is the main medium because it can be accurately modelled and is light and durable. The skin is then positioned over the 'armature' and carefully stitched up. Finally, the facial features are modelled and artificial glass eyes are inserted, perhaps with some additional art work to bring the 'mount' to life. Perhaps the greatest attribute needed by the taxidermist is a genuine love of natural history. Shown here is part of the musculature of a lion being modelled by binding wood-wool onto an armature.

The end of a day at the Museum. Shortly before six, with not too much solemnity, a handbell is rung. Reluctant children have their coats put on, with promises of a longer visit next time. One by one the lights go out and the Main Hall becomes a dark cavern, echoing to the steps of warders as they report each Gallery cleared. Staff hurry for the homeward train, while *Diplodocus*, suddenly free from wires and metal supports, seems poised to take an evening stroll. Outside, its terracotta glowing in the floodlights, the façade looks every inch a storehouse for the 'Wonders of Creation'.

Acknowledgements

The stereoscan photographs were supplied by the Museum's Electron Microscope Unit and the photomicrographs were undertaken by Peter York of the Photographic Unit. A few pictures could not be taken from Museum material and the following are gratefully acknowledged for supplying their own excellent photographs, for allowing us to use previously published material, or for loaning specimens:

Dr W. Sands (Centre for Overseas Pest Research)
Dr D. Williams (Commonwealth Institute of Entomology)
P. J. Chimonides (Museum staff)
Dr D. George (Museum staff)
Controller of Her Majesty's Stationary Office and the Director, Royal Botanic Gardens, Kew
Dr S. E. O. Meredith and the Liverpool School of Tropical Medicine
British Museum, Bloomsbury
Winchester Research Unit, Winchester

The subjects in this book were the result of discussions throughout the Museum and were mostly suggested by my colleagues in the various Departments, who most kindly supplied material for the text, checked my version of it, and made available the relevant material for photography. To them I owe considerable gratitude, not only for the trouble that they took, but for often introducing me to specimens, stories, and aspects of natural history that I would otherwise have missed. My very warm thanks go to:

Dr P. J. Andrews
Dr E. N. Arnold
Dr H. W. Ball
G. A. Ballance
A. Bedser
A. C. Bishop
Dr I. R. Bishop
Dr R. L. Blackman
A. F. Blake
H. A. Buckley
J. F. M. Cannon
A. O. Chater
Dr A. M. Clark
Dr G. C. S. Clarke
D. Claugher
C. A. Comben
P. L. Cook
Dr G. B. Corbet
A. G. Couper
A. J . Criddle
Dr R. W. Crosskey
R. Croucher
J. A. Diment
P. I. Edwards (retired)
Dr G. F. Elliott
P. G. Embrey
Dr A. A. Fincham
Dr P. L. Forey
Dr P. Freeman
I. C. G. Galbraith
Dr J. D. George
Dr D. I. Gibson
P. Gilbert
R. P. D. Goodwin
Dr A. L. Graham
Dr P. H. Greenwood
R. F. Hale
P. M. Hammond

R. H. Harris
Dr R. G. Harvey
C. R. Hill
Dr M. Hills
Dr M. K. Howarth
G. J. Howes
Dr R. W. Ingle
P. W. James
Dr R. P. Jefferies
Dr J. Jewell
Dr B. W. Kamill
C. B. Keates
D. N. Lewis
G. A. Mathews
Dr A. C. Milner
D. T. Moore
L. J. Moore
Dr P. B. Mordan
P. H. Napier
C. P. Nuttall
C. G. Ogden
C. J. Owen
T. B. B. Paddock
R. J. Pankhurst
R. J. Parsons
Dr C. Patterson
R. D. Pope
J. Quinlan
Dr D. R. Ragge
Dr B. R. Rosen
M. J. Rowlands
Dr W. A. Sands (Centre for Overseas Pest Research)
Dr K. S. O. Sattler
Dr J. G. Sheals
M. C. Sheldrick
F. G. A. M. Smit

Dr W. T. Stearn (retired)
S. M. K. Stone
Dr J. D. Taylor
R. I. Vane-Wright
A. R. Vickery
F. R. Wanless
A. Watson
K. M. Way
Dr P. E. S. Whalley
A. C. Wheeler
Dr G. B. White (London School
    of Hygiene and Tropical Medicine)
Dr J. E. P. Whittaker
S. Whybrow
Dr D. J. Williams (Commonwealth
    Institute of Entomology)
Dr A. R. Woolley
Dr C. A. Wright

The text was most critically read and numerous facts rechecked by my colleague
Dr Klaus Sattler while making the German translation. I am indebted to him for
many corrections and helpful suggestions.